DEVELOPMENTS IN MINERAL PROCESSING 17

SOLUTION CHEMISTRY: MINERALS AND REAGENTS

DEVELOPMENTS IN **MINERAL PROCESSING 17**

SOLUTION CHEMISTRY: MINERALS AND REAGENTS

P. SOMASUNDARAN
COLUMBIA UNIVERSITY, NEW YORK, USA

WANG, DIANZUO
CENTRAL-SOUTH UNIVERSITY, CHANGSHA, CHINA

SERIES EDITOR: B.A. WILLS

ELSEVIER

Amsterdam • Boston • Heidelberg • London • New York • Oxford
Paris • San Diego • San Francisco • Singapore • Sydney • Tokyo

Elsevier
Radarweg 29, PO Box 211, 1000 AE Amsterdam, The Netherlands
The Boulevard, Langford Lane, Kidlington, Oxford OX5 1GB, UK

First edition 2006

Library of Congress Cataloging-in-Publication Data
A catalog record for this book is available from the Library of Congress

British Library Cataloguing in Publication Data
A catalogue record for this book is available from the British Library

ISBN-13: 978-0-444-52059-3
ISBN-10: 0-444-52059-7
ISBN: 0167-4528

For information on all Elsevier publications
visit our website at books.elsevier.com

Transferred to Digital Printing in 2010

Contents

Appendices

Preface

Surfactants have been used for many industrial processes such as flotation, enhanced oil recovery, soil remediation and cleansing. Flotation technology itself has been used in industry since the end of 19th century, and even today it is an important method for mineral processing and its application range is expanding to other areas. Thus it has been used in the treatment of wastewater, waste materials from industries, separation and recycling of municipal wastes and some unit processes of chemical engineering. The efficiency of all these operations depends primarily on the interactions among surfactants, solids and media. In addition, various chemical additives such as polymers and dispersants also introduce another dimension to the solution chemistry of the system. These interactions in turn determine the wettability and dispersion behavior of solids as well as stability of foams and emulsions in the system.

The basic interfacial process in flotation is selective hydrophobization (or lepophilization) and hydrophilization of particulate matter. The role of the solution chemistry is very important in flotation as it is determined by the dissolution behavior of mineral particles in the aqueous solution (pulp) and subsequent dissociation, hydrolysis and precipitation of the soluble species; the dissolution, association, dispersion and emulsion behavior of various flotation reagents in the pulp and interactions among reagents with both soluble and surface species of minerals. The efficiency of flotation and separation of mineral particles and consumption of reagents are thus controlled by the solution chemistry of the pulp. As other processes such as oil displacement are also governed by such interactions and in turn by the wettability of the solid surface, the study of solution chemistry of surfactant/mineral/additive systems become very important for the development of many technologies.

In this book, the fundamentals of solution chemistry of mineral/surfactant systems and important calculations involved are systematically introduced and the influence of relevant physico-chemical conditions are discussed. Importantly, because solution chemistry is closely related to the molecular structure of the flotation reagents, the characteristics of such structure is also discussed in detail.

Professor Dianzuo Wang of Central-south University (Changsha, China) and Professor P. Somasundaran of Columbia University (New York City, U.S.) would like to acknowledge the help of many students and associates in the mineral processing and applied surface chemistry laboratories of the two universities: Professor Hu Yuehua (Changsha, China) who revised some chapters of this book, Dr. Liping Xiao who organized materials for some chapters, Dr. S. Kumar, Dr. A. Lou, Dr. Dong Qinghai, Dr. L. Zhang and Ms. Jing Wang who helped in sorting out and editing many versions of the book.

Professor Wang Dianzuo
Changsha
Professor P. Somasundaran
New York

Chapter 1

Introduction

1.1. BRIEF REVIEW OF THE SOLUTION CHEMISTRY OF REAGENT/MINERAL SYSTEMS

Physico-chemical interactions in aqueous solutions play a governing role in determining the attachment of air bubbles or oil droplets to particles and hence their behavior in processes such as flotation. These interactions are in turn determined by the properties of a number of system components including surface active flotation reagents such as collectors, frothers and other inorganic/organic/polymeric additives and constituent minerals/solids particles/and air bubbles/oil droplets. Most of these components bring about their own characteristic chemical interactions which often become exceedingly complex in the presence of one another. Relevant major interactions include self association, dissociation and precipitation of surfactants, dissolution of minerals, hydrolysis/complexation/precipitation of different species and the reactions of surfactants with other species in the solution. In addition to the adsorption of various species, several reactions similar to the above bulk reactions can occur also at the solid/liquid interface and they will have a predominant role in determining the ultimate surfactant adsorption and thus flotation, flocculation and other colloidal behavior. Specifically, chemical alterations at the particle surface resulting from adsorption or surface precipitation of surfactants and the orientation of these species at the solid/liquid interface, will determine the hydrophobicity or otherwise of the particles. Attachment of hydrophobic particles to air bubbles, on the other hand, will depend additionally upon the orientation and dynamics of the species adsorbed at the liquid–air interface, which is also governed by the solution chemistry of the reagents. Thus, the solution chemistry as well as mineral surface chemistry plays a fundamental role in the processes such as flotation and flocculation.

Since the 1920's, a great number of water-soluble synthetic surfactants have been developed and commercially used in the mineral processing industry over the years. Flotation process has become a cost effective and highly efficient industrial technology. Flotation theories were also developed along with the commercialization of the flotation process. These theories expounded by researchers were based (mostly) on the application of fundamental chemical knowledge to explain the interactions between minerals and reagents involved in flotation. It was Taggart (1945) who for the first time propounded in 1930 a chemical reaction hypothesis: the solubility product theory. He suggested that the flotation of mineral particles depends upon chemical reactions of flotation reagents with the mineral surface. Smaller the solubility product of the compounds formed by reagents and the metallic ions of a mineral, stronger is the interaction between the reagents and the minerals and better is the flotation. A typical example of the application of this principle is ethyl xanthate/galena (PbS)/sphalerite (ZnS) system. Ethyl xanthate can float galena but not sphalerite. Based on the solubility products of $K_{\mathrm{spZnEX_2}}/K_{\mathrm{spPbEX_2}} = 4.9\times10^{-9}/1.7\times10^{-17} =$

2.8×10^8, ethyl xanthate has a stronger interaction with galena than sphalerite and causes better flotation of it.

A significant development in this regard was the correlation of the solubility products of a series of heavy metal-ethyl xanthate salts with the floatability of corresponding sulfide minerals by (Kakovsky, 1980). He found the decrease in the order of the solubility product of sulfide minerals to be in line with the increase in the order of their floatability. From exchange reactions of lead-diethyl xanthate, the well-known Barsky equation can be derived:

$$PbX_2 + 2OH^- = Pb(OH)_2 + 2X^- \tag{1.1}$$

$$K' = [X^-]^2/[OH^-]^2 = K_{spPbX_2}/K_{spPb(OH)_2} = \text{const.} \tag{1.2}$$

Thus,

$$[X^-]/[OH^-] = (K')^{1/2} = \text{const.}$$

However, Gaudin (1932) found $[X^-]/[OH^-]$ not to be a constant under critical flotation conditions. Instead, the governing equation was proposed to be:

$$[X^-] = m[OH^-]^y, \tag{1.3}$$

where m and y are constants. Since the 1950's, it has been recognized by many researchers that complex flotation phenomena could not be so easily explained. du Reitz (1976), Marabini (1984) and Hu et al. (1996) tackled this problem differently by developing species distribution diagrams (such as $\log C$–pH, and ΔG^0–pH diagrams) by considering multi-chemical equilibria of hydrolysis of cations, proton-addition of anions, complexation of metallic ions with reagents, etc.

In the case of long chain surfactants used as collectors, a significant correlation between their adsorption in the form of aggregates and flotation of minerals with them was established in the 1960's by Somasundaran et al. (1964). Somasundaran et al. (1976, 1985) developed the relevant dissociation and aggregation equilibria for flotation reagents in the solution as well as at the mineral–solution interface. Importantly, the "ion–molecule complexes" phenomenon was proposed later by Somasundaran (1976), Hanna and Somasundaran (1976) to account for the flotation maximum exhibited at certain pH values by hydrolyzable surfactants.

Flotation of sparingly soluble minerals such as calcite has not been well understood. Recently, Somasundaran et al. (1985, 1991) and Hu et al. (1995) have done some systematic work on the dissolution equilibria of salt-type minerals in solutions and their effects on the selectivity in flotation. Thus they found that surface conversion of mineral surfaces and precipitation on surfaces of minerals due to interactions among dissolved mineral species and the reagents are the major mechanisms that should be considered in flotation systems of sparingly-soluble minerals.

It can be concluded from the above discussion that chemical equilibria in solutions play a critical role in determining the surface chemical properties of minerals and the behavior of flotation systems. More and more emphasis has been laid by researchers on the solution chemistry of flotation in recent years (Hu et al., 2000; Celik et al., 2002; Cetin, 2002). The objective of this book is to summarize the work that has been done in the last several decades on solution chemistry, describe various solution chemical equilibria systematically and point out future trends in this area.

1.2. OUTLINE OF THE SOLUTION CHEMISTRY OF FLOTATION

Following four aspects of the Solution chemistry of flotation are considered:

(1) Solution equilibria of flotation reagents. Solution equilibria of various surfactants such as collectors, frothers, modifiers, and flocculants, encompass acid–base equilibria, dissociation equilibria, association equilibria and polymer equilibria. Using these equilibria and also various parameters such as the pH of the solutions and the critical pH of the dissolution of insoluble flotagents, pK_a can be calculated and the state of active flotation reagent species in solution predicted.
(2) Mineral–solution equilibria. Mineral–solution equilibria describe the dissolution of minerals, and exert a critical role in relevant phenomena such as hydration of dissolved mineral species, formation of precipitates and their electrokinetic behavior. Important parameters like concentration of various dissolved mineral species/pH/onset of precipitation and point of zero charge can be calculated from mineral–solution equilibria. All these factors are helpful for determining optimum conditions for flotation.
(3) Mineral-surfactant equilibria. Mineral–surfactant equilibria determine adsorption of various surfactants on minerals, hemimicellization, interactions among dissolved mineral species and reagents and electrochemical interactions. The optimum conditions for mineral-surfactant interactions as well as flotation can be calculated from data for such parameters as the solubility product and complex formation constants and species distribution diagrams with the helps of plots of $\log C$–pH, ΔG^0–pH, $\log \beta'_n$, E_h–pH, etc.
(4) Structure-property relationship of surfactants. Flotation reagents are essentially of three groups: hydrophilic, mineralophilic and hydrophobic. The properties of flotation reagents such as collectors, frothers, depressants and flocculants depend mainly upon the structure of functional groups in the reagent molecule. Data on a number of structural effects (induction, conjugation and steric factor) and quantitative criteria (group electronegativity, HLB and solubility) can be used to determine the structure–property relationships for various reagents. Such information can in turn be used to develop molecular design of useful flotation reagents for given mineral systems.

REFERENCES

Celik, M., Hancer, M., Miller, J.D., 2002. J. Coll. Inter. Sci. 256 (1), 121–131.

Cetin, K., 2002. Colloids and Surfaces A 210 (1), 23–31.
du Reitz, C., 1976. In: Proceed. 11th Inter. Miner. Process. Congress. Cagliari, Italy, pp. 375–403.
Gaudin, A.M., 1932. Flotation. McGraw-Hill Book Co., New York, pp. 552.
Hanna, H.S., Somasundaran, P., 1976. In: Fuerstenau, M.C. (Ed.), Flotation—Gaudin Memorial Volume, vol. 1. AIME, p. 197.
Hu, Y.H., Luo, L., Qiu, G.Zh., Wang, D.Z., 1995. Transactions of Nonferrous Metals Society of China 5, 26–30.
Hu, Y.H., Qiu, G.Zh., Xu, J., Wang, D.Z., 1996. J. of Mining and Metallurgy 1, 28–33.
Hu, Y.H., Liu, X.W., Qiu, G.Zh., 2000. Mining and Metallurgical Engineering 20 (2), 11–14.
Kakovsky, I.A., Schkaleva, P.H., 1980. Theory Basis and Control of Flotation Process (in Russian). Science Press, Moscow, pp. 94–105.
Marabini, A.M., et al., 1984. In: Jones, M.J., Oblatt, R. (Eds.), Reagents in Mineral Industry, pp. 125–136.
Somasundaran, P., 1976. Inter. J. Miner. Process. 3, 35–40.
Somasundaran, P., Healy, T.W., Feurstenau, D.W., 1964. J. Phys. Chem. (12), 3562–3566.
Somasundaran, P., Amankonah, J.O., Ananthapadmanabhan, K.P., 1985. Colloids & Surfaces 15, 309–333.
Somasundaran, P., Liping, X., Dianzuo, W., 1991. Mineral & Metallurgical Process 8 (3), 115–121.
Taggart, A.F., 1945. Handbook of Mineral Dressing. John Wiley & Sons, New York.

Chapter 2

Solution equilibria of surfactants

Reagents used in flotation, collectors, frothers, depressants, flocculants and inorganic modifiers can interact with each other in the flotation pulp and at the mineral–solution interface. The chemical equilibria involved in these interactions and the nature of the products will have a significant effect on their adsorption and the resultant flotation processes.

Most of the theories on interactions of surfactants with minerals are closely related to their solution chemistry. For example, the ion-exchange adsorption theory proposed by Gaudin (1932, 1934) and Wark (1938) and the molecular adsorption theory proposed by Cook and Nixon (1950) are based on the dissociation equilibria and states of the collectors in water. More recently, Somasundaran (1976) observed that ion-molecule complexes of long-chain surfactants in flotation systems can have high surface activity depending upon the association equilibria of the surfactants in solutions (Ananthapadmanabhan et al., 1979; Kulkarni and Somasundaran, 1980). Also the cationic flotation behavior of salt type minerals is closely related to the formation of alkyl amine salt (Hu and Wang, 1990). In this chapter, solution equilibria of reagents relevant to selected flotation systems are examined.

2.1. ACID–BASE EQUILIBRIA IN DILUTE FLOTATION SOLUTIONS

2.1.1. Solution equilibria equivalents

According to Sugha and Kotrly (1972), it was Arrhenius who, toward the end of the 19th century, defined for the first time substances which release H^+ ions in their aqueous solutions as acids (AH) and those which release OH^- as a base (BOH):

$$AH = A^- + H^+; \qquad BOH = B^+ + OH^- \qquad (2.1)$$

In 1923, Lewis (Sugha and Kotrly, 1972) proposed an acid–base theory in which the electron acceptor was defined as a an acid and the electron donor as a base. Most reagents except neutral oils can be considered as an acid or a base. For example, cationic activators such as Cu^{2+}, Pb^{2+}, Ca^{2+} are acids and collectors such as R–COO$^-$, R–O–CSS$^-$ are bases. Solution equilibria of these reagents can therefore be derived from their acid–base equilibria. Four basic formulae required for equilibrium calculations are given below with Na_2CO_3 as the example:

(1) Proton transfer equivalent (PTE). Protons can be transferred from various compounds in solutions according to the following dissociations.

$$Na_2CO_3 + 2H_2O = H_2CO_3 + 2NaOH \qquad (2.2)$$

$$H_2CO_3 = H^+ + HCO_3^- \qquad K_{a1} = [H^+][HCO_3^-]/[H_2CO_3] \qquad (2.3)$$

$$HCO_3^- = H^+ + CO_3^{2-} \qquad K_{a2} = [H^+][CO_3^{2-}]/[HCO_3^-] \tag{2.4}$$

$$H_2O = H^+ + OH^- \qquad K_w = [H^+][OH^-] \tag{2.5}$$

where K_{a1}, K_{a2} and K_w are the first and second dissociation constants of the carbonic acid and the ionic product of water, respectively.

(2) Mass balance equivalent (MBE). The total concentration of a reagent should be equal to the sum of various species generated by it

$$C_{Na_2CO_3} = [H_2CO_3] + [HCO_3^-] + [CO_3^{2-}] \tag{2.6}$$

$$C_{Na_2CO_3} = [Na^+]/2 \tag{2.7}$$

(3) Charge balance equivalent (CBE). The total concentration of positively charged species should be equal to that of the negatively charged species, maintaining electrical neutrality in the solution.

$$[H^+] + [Na^+] = [HCO_3^-] + 2[CO_3^{2-}] + [OH^-] \tag{2.8}$$

(4) Proton balance equivalent (PBE). The number of protons accepted by the original species should be equal to that given by other original species. For PBE of HCO_3^- obtained from the combination of Eqs. (2.6), (2.7) and (2.8).

$$[H_2CO_3] + [H^+] = [OH^-] + [CO_3^{2-}] \tag{2.9}$$

Similarly, PBE of H_2CO_3 and H_2O, CO_3^{2-} and H_2O are

$$[H^+] = [HCO_3^-] + [CO_3^{2-}] + [OH^-] \tag{2.10}$$

$$[H^+] + [HCO_3^-] + [H_2CO_3] = [OH^-] \tag{2.11}$$

2.1.2. Acid–base equilibrium calculation

Surfactants as acids or bases can undergo hydrolysis or dissociation reactions in aqueous solutions and change the pH of the solution. This change can further affect the interactions among reagents and minerals and between reagent species themselves. It is therefore useful to have a knowledge of the pH of solutions of various types of reagents under different conditions on the basis of the above acid–base equilibria.

2.1.2.1. *Strong acid and strong base type of reagents*

Only a few reagents such as pH modifiers HCl and H_2SO_4 and collectors such as alkyl sulfonic acid R–SO_3 H or sulfonic acid R–O–SO_3H, belong to the strong acid type and pH modifiers such as NaOH and collectors like quaternary ammonium R–$N(CH_3)_3^+$ belong to the strong base type. The pH of solutions of this type of reagents can be determined from the concentration of hydrogen ions or hydroxyl ions in dilute solutions:

$$C_A = [H^+] \quad \text{or} \quad C_B = [OH^-], \tag{2.12}$$

where C_A and C_B are concentrations of acid and base, respectively.

2.1.2.2. *Mono weak acid-weak base weak type of reagents*

For a weak monoatomic acid, HA, relevant equilibria in solutions are:

$$\text{PBE:} \quad HA = H^+ + A^- \qquad K_a = [H^+][A^-]/[HA] \tag{2.13}$$

$$\text{MBE:} \quad C_A = [A^-] + [HA] \tag{2.14}$$

Substitute Eq. (2.13) into Eq. (2.14)

$$[A^-] + [HA] = [A^-] + [H^+][A^-]/K_a \approx [H^+] + [H^+]^2/K_a$$

$$[H^+]^2 + K_a[H^+] - K_aC_A = 0$$

$$[H^+] = -K_a/2 + (K_a^2/4 + K_aC_A)^{1/2} \tag{2.15}$$

For a weak mono base

$$\text{PBE:} \quad BH^+ = B + H^+ \tag{2.16}$$

$$K_a = [H^+][B]/[BH^+] = [H^+][OH^-][B]/[BH^+][OH^-] = K_w/K_b, \tag{2.17a}$$

where $K_w = [H^+][OH^-]$ and $K_b = [BH^+][OH^-]/[B]$ is the equilibrium base-association constant for the reaction:

$$B + H_2O = BH^+ + OH^- \tag{2.17b}$$

From MBE:

$$C_B = [B] + [BH^+] = [B](1 + [BH^+]/[B]) \tag{2.18}$$

From Eq. (2.17b), $[B] = [BH^+][OH^-]/K_b \approx [OH^-]^2/K_b$

$$[BH^+]/[B] = K_b/[OH^-]$$

Thus

$$C_B = [OH^-]^2(1 + K_b/[OH^-])/K_b \tag{2.19a}$$

$$[OH^-]^2 + K_b[OH^-] - K_bC_B = 0 \tag{2.19b}$$

$$[OH^-] = -K_b/2 + (K_b^2/4 + K_bC_B)^{1/2} \tag{2.19c}$$

or

$$[OH^-] = -K_w/2K_a + (K_w^2/4K_a^2 + C_BK_w/K_a)^{1/2} \tag{2.19d}$$

pH values of solutions of commonly used weak mono-acid and mono-base reagents at different concentrations are calculated using Eqs. (2.15) and (2.19) and the results are given in Table 2.1. As can be seen from Table 2.1, solutions of xanthate and oleic acid are acidic whereas those of amine are basic in the concentration range (10^{-5}–10^{-3} mol/l) used in flotation processes.

Table 2.1
pH values of solutions of common weak mono-acid and mono-base reagents at different concentrations

Flotation reagent	K_a	Concentration (mol/l)						
		10^{-1}	10^{-2}	10^{-3}	5×10^{-4}	10^{-4}	5×10^{-5}	10^{-5}
Xanthate acid	10^{-5}	3.0	3.5	4.0	4.2	4.6	4.7	5.2
Oleic acid	10^{-5}	3.5	4.0	4.5	4.7	5.0		5.6
Hydrocyanic acid	$10^{-9.21}$	5.1	5.6	6.1		6.6		
Dodecyl amine	$10^{-10.63}$	11.8	11.3	10.7		9.9		8.3

Table 2.2
pH values of solutions of common weak mono-acid and strong mono-base reagents at different concentrations

Flotation reagent	K_a	Concentration (mol/l)					
		10^{-1}	10^{-2}	10^{-3}	10^{-4}	10^{-5}	10^{-6}
Sodium ethylxanthate	10^{-5}	9.0	8.5	8.0	7.5	7.2	7.0
Sodium dithiophosphate	2.3×10^{-5}	8.8	8.3	7.8	7.4	7.1	7.0
Sodium oleate	10^{-6}	9.5	9.0	8.5	8.0	7.5	7.2
Sodium octyl hydroxamate	10^{-9}	11.0	10.5	10.0	9.5	9.0	8.5
Sodium cyanide	$10^{-9.21}$	11.1	10.6	10.1	9.6	9.1	8.6

2.1.2.3. *Monoatomic strong base–weak acid type reagents*

For a monoatomic strong base–weak acid salt

$$\text{PBE:} \qquad [H^+] + [HA] = [OH^-] \tag{2.20}$$

$$[H^+] + [H^+][A^-]/K_a = K_w/[H^+]$$

$$[H^+]^2(1 + [A^-]/K_a) = K_w$$

Thus

$$[H^+] = \{K_a K_w/(K_a + [A^-])\}^{1/2} \tag{2.21a}$$

For strong base and weak acid salt $[A] \approx C_{BA}$

$$[H^+] = [K_a K_w/(K_a + C_{BA})]^{1/2} \tag{2.21b}$$

pH values of solutions obtained with commonly used strong base–weak acid type reagents are given in Table 2.2. It can be seen that the solutions of these reagents exhibit different basic pH values depending on their concentration.

2.1.2.4. *Weak multiatomic acid type reagents*

For a weak multi-acid type of reagent, H_nA, the proton transfer equilibrium can be considered to proceed in multiple steps:

$$A^{n-} + H^+ = HA^{n-1}, \qquad K_1^H = [HA^{n-1}]/[A^{n-}][H^+] = \beta_1^H$$

$HA^{n-1} + H^+ = H_2A^{n-2}, \qquad K_2^H = [H_2A^{n-2}]/[H^+][HA^{n-1}],$

$\beta_2^H = [H_2A^{n-2}]/[H^+]^2[A^{n-}]$

...

$H_{n-1}A^- + H^+ = H_nA, \qquad K_n^H = [H_nA]/[H_1A^-][H^+],$

$$\beta_n^H = \frac{[H_nA]}{[H^+]^n[A]} = \prod_{i=1}^{n} K_i^H \tag{2.22}$$

where $K_1^H, K_2^H, \ldots, K_n^H$ are proton-addition constants for different steps and related to association constants as below:

$K_n^H = 1/K_{a1}, \ldots, K_2^H = 1/K_{a,n-1}, K_1^H = 1/K_{a,n},$

$\beta_1^H, \beta_2^H, \ldots, \beta_n^H$ are cumulative proton-addition constants.

Assuming [A], [A′] to be the concentrations of individual A and total A.

$$[A'] = [A^{n-}] + [HA^{n-1}] + \cdots + [H_nA] = [A^{n-}] + \beta_1^H[A^{n-}][H^+] + \cdots + \beta_n^H[A^{n-}][H^+]^n \tag{2.23}$$

When only a multiatomic weak-acid is present in solution, PBE will be:

$$[H^+] = [H_{n-1}A^-] + 2[H_{n-2}A^{2-}] + \cdots + (n-1)[HA^{n-1}] + n[A^{n-}] + [OH^-] \tag{2.24a}$$

Since $[H_nA]$ and $[H_{n-1}A^-]$ are the major components in multiatomic weak-acid solution, $[OH^-]$ can be neglected. The PBE is then simplified as:

$$[H^+] \approx [H_{n-1}A^-] = \beta_{n-1}^H[A^{n-}][H^+]^{n-1} \tag{2.24b}$$

Then

$$C_{H_nA} \approx [H_{n-1}A^-] + [H_nA] = \beta_{n-1}^H[A^{n-}][H^+]^{n-1} + \beta_n^H[A^{n-}][H^+]^n \tag{2.25}$$

$$C_{H_nA}/[H^+] = \{\beta_{n-1}^H[A^{n-}][H^+]^{n-1} + \beta_n^H[A^{n-}[H^+]^n\}/\beta_{n-1}^H[A^{n-}][H^+]^{n-1} = 1 + K_n^H[H^+]$$

$$K_n^H[H^+]^2 + [H^+] - C_{H_nA} = 0 \tag{2.26}$$

pH values of some multiatomic weak acid flotation reagents are calculated using Eq. (2.26) and given in Table 2.3. It can be seen from the table that the multiatomic acid flotation reagents usually show strong acidity in solutions. A certain amount of base has to be added as collectors, depressants or dispersants in neutral or basic medium when this type of reagents are used.

2.1.2.5. *Multiatomic weak acid–strong base type of reagents*

For a multiatomic weak acid–strong base salt, B_nA, PBE of A will be:

$$[H^+] + [HA^{n-1}] + 2[H_2A^{n-2}] + \cdots + (n-1)[H_{n-1}A^-] = OH^- \tag{2.27}$$

Table 2.3
pH Values of solutions of some polyacid reagents at different concentrations

Flotation reagents	K_n^H	Concentration (mol/l)				
		10^{-1}	10^{-2}	10^{-3}	10^{-4}	10^{-5}
Methyl-benzenearsonic acid	5×10^3	2.36	2.88	3.45	4.14	5.00
Phosphoric acid	$10^{2.15}$	1.6	2.2	3.0	3.91	4.3
Citric acid	$10^{3.13}$	2.1	2.6	3.2	4.0	5.0
Tartaric acid	$10^{3.93}$	2.5	3.0	3.5	4.2	5.0
Oxalic acid	$10^{1.25}$	1.3	2.1	3.0	4.0	5.0

Table 2.4
pH values of solutions of some salts with multi-weak acid–strong base reagents at different concentrations

Flotation reagents	K_1^H for acid	Concentration (mol/l)					
		10^{-1}	10^{-2}	10^{-3}	10^{-4}	10^{-5}	10^{-6}
Sodium phosphate	$10^{12.35}$	12.6	11.9	11.0	10.0	9.0	8.0
Sodium carbonate	$10^{9.57}$	11.3	10.8	10.2	9.7	8.9	
Sodium sulfide	$10^{13.8}$	12.9	12.0	11.0	10.0	9.0	8.0
Sodium oxalate	$10^{4.28}$	8.6	8.1	7.6	7.1		
Sodium metasilicate	$10^{12.0}$	12.4	11.8	11.0	10.0	9.0	8.0

Since $[A^{n-}]$ and [HA] are major species and $[H^+]$ can be ignored, PBE can be simplified as:

$$[HA^{n-1}] = [OH^-] \tag{2.28}$$

On the other hand,

$$C_{B_nA} = [A^{n-}] + [HA^{n-1}] = [HA^{n-1}]/K_1^H[H^+] + [OH^-] = [OH^-]/K_1^H[H^+] + [OH^-]$$

$$[OH^-]^2 + K_1^H K_w[OH^-] - C_{B_nA} = 0 \tag{2.29}$$

or

$$[H^+]^2 - K_w[H^+]/C_{B_nA} - K_w/K_1^H C_{B_nA} = 0$$

Thus

$$[OH^-] = -K_1^H K_w/2 + [(K_1^H K_w)^2/4 + C_{B_nA}]^{1/2} \tag{2.30a}$$

$$[H^+] = K_w/2C_{B_nA} + [K_w^2/4C_{B_nA}^2 + K_w/K_1^H C_{B_nA}]^{1/2} \tag{2.30b}$$

pH values of solutions of multiatomic weak–acid strong base flotation reagents calculated using Eq. (2.30b) are given in Table 2.4. These flotation reagents can be used as depressants, dispersants and activators. They can also be used as pH modifiers when flotation is carried out in basic medium.

2.2. DISSOCIATION AND HYDROLYSIS EQUILIBRIA OF SHORT-CHAIN REAGENTS

2.2.1. Anionic flotagents

In the case of monoatomic weak acid–strong base salt (NaA), hydrolysis takes place first and then dissociation:

$$NaA + H_2O = HA + NaOH \quad (2.31)$$

$$HA = H^+ + A^- \qquad K_a = [H^+][A^-]/[HA] \quad (2.32)$$

or in the log form:

$$pH - pK_a = \log[A^-]/[HA] \quad (2.33)$$

The K_a values of commonly used anionic flotation reagents are shown in Appendix A. From Eq. (2.33), the effective pH range of various flotation reagents (for electrostatic force) can be predicted from the pK_a of flotation reagents and the point of zero charge (PZC) of the mineral.

(a) If anionic flotation reagents interact with minerals only through electrostatic forces, the effective pH range of weak monatomic acid type of flotation reagents will be:

$$pK_a < pH < PZC \quad (2.34)$$

(b) If anionic flotation reagents interact with minerals mainly through molecular adsorption, the effective pH range will be:

$$pH < pK_a, \qquad [HA] > [A^-] \quad (2.35)$$

2.2.2. Cationic flotation reagents

The dissociation of cationic flotagents may be classified into two kinds: acid dissociation and base dissociation. Consider amine as an example.

Acid dissociation:

$$RNH_3^+ = RNH_{2(aq)} + H^+$$

$$K_a = [H^+][RNH_{2(aq)}]/[RNH_3^+] = 10^{-10.63} \quad (2.36)$$

Base dissociation:

$$RNH_{2(aq)} + H_2O = RNH_3^+ + OH^-$$

$$K_b = [OH^-][RNH_3^+]/[RNH_{2(aq)}] = 4.3 \times 10^{-4}, \quad (2.37)$$

where $K_b = K_w/K_a$, log form of Eq. (2.36) is

$$pH - pK_a = \log[RNH_{2(aq)}]/[RNH_3^+] \quad (2.38)$$

K_a values of commonly used cationic flotation reagents are shown in Appendix A. The effective pH range for a system can be predicted from Eq. (2.38), pK_a of flotation reagents and point of zero charge (PZC) of minerals.

Table 2.5
pH_0 values for some amphoteric reagents

Flotation reagents		pH_0
N-Hexadecyl-aminonacetic acid	$(C_{16}H_{33}NHCH_2COOH)$	4.5
N-Coconut oil-β-amino-isobutyric acid	$(RNHCH(CH_3)CH_2COOH)$	4.1
N-Dodecyl-β-amino-propionic acid	$(C_{12}H_{25}NHCH_2CH_2COOH)$	4.3
N-Dodecyl-β-nitrilo-dipropionic acid	$(C_{12}H_{25}N(CH_2CH_2COOH)_2)$	3.7
N-Tetradecyl-amino-ethylsulfonic acid	$(C_{14}H_{29}NHCH_2CH_2SO_3H)$	1.0
Stearic aminosulfonic acid		6.3–6.6
Oleic aminosulfonic acid		6.3–6.6

(a) If cationic flotation reagents interact with minerals such as quartz, fluospar and rutile only through electrostatic force, the effective pH range will be:

$$pK_a > pH > PZC \tag{2.39}$$

(b) If cationic flotation reagents interact with minerals to form molecular complexes, $Me(RNH_2)^{n+}$, the effective pH rage will be:

$$pH > pK_a \tag{2.40}$$

2.2.3. Amphoteric flotation reagents

Amphoteric flotagents can behave both like anionic and cationic molecules depending upon pH of the solution. Consider amino acid as an example.

Anionic type in basic solution:

$$RNHCH(CH_3)CH_2COOH + OH^- = RNHCH(CH_3)CHCOO^- + H_2O \tag{2.41}$$

Cationic type in acidic solution:

$$RNHCH(CH_3)CH_2COOH + H^+ = RNH_2^+CH(CH_3)CH_2COOH \tag{2.42}$$

Around the pH at which the anionic and cationic ions are in equilibrium called point of zero charge (PZC), pH_0, the behavior of flotation reagents on minerals will reverse.

(a) When $pH_0 < PZC$, the effective range of electrostatic interaction on minerals will be:

$$pH_0 < pH < PZC \tag{2.43}$$

(b) When $pH > PZC$, the effective range will be:

$$pH_0 > pH > PZC \tag{2.44}$$

The pH_0 values of some amphoteric flotation reagents are shown in Table 2.5 (Nemethy and Scheraga, 1962). The PZC values of various minerals are given in Chapter 3. The relationship between flotation behavior of salt type minerals and solution equilibria of alkyl amino phosphoric acid has also been reported (Hu et al., 2003).

2.3. DIAGRAM APPROACH FOR SOLUTION EQUILIBRIA OF FLOTATION REAGENTS

To demonstrate correlation of flotation with complicated solution equilibria of multiatomic acids and bases and salts such as Na_2S, Na_2CO_3, and $H_2C_2O_4$, different diagrammatic analyses are used.

2.3.1. ϕ–pH diagram

From Eq. (2.22), additional reaction coefficients can be obtained as:

$$a_A = [A']/[A] = 1 + \beta_1^H[H] + \beta_2^H[H]^2 + \cdots + \beta_n^H[H]^n \quad (2.45)$$

Fractions of various species in solution at total concentration of [A′] will respectively be:

$$[A]/[A'], \qquad [HA]/[A'], \qquad [H_2A]/[A'], \qquad [H_nA]/[A'],$$

$\phi_0, \phi_1, \ldots, \phi_n$ are distribution coefficients of various species as a function of pH, representing their concentration fractions.

$$\phi_0 = [A]/[A'] = 1/a_A = 1/(1 + \beta_1^H[H] + \beta_2^H[H]^2 + \cdots + \beta_n^H[H]^n)$$
$$\phi_1 = [HA]/[A'] = K_1^H[H][A]/[A'] = K_1^H\phi_0[H]$$
$$\phi_2 = [H_2A]/[A'] = \beta_2^H[H]^2[A]/[A'] = \beta_2^H\phi_0[H]^2$$
$$\ldots$$
$$\phi_n = [H_nA]/[A]' = \beta_n^H[H]^n[A]/[A'] = \beta_n^H\phi_0[H]^n. \quad (2.46)$$

The species distribution diagrams (ϕ–pH) for flotation reagents can be calculated from proton addition constants given in Appendices A and B. Some examples are given below.

2.3.1.1. *ϕ–pH diagram and flotation effect of Na_2S*

Na_2S is a sulfidizing agent for copper, lead and zinc oxides and depressant for sulfide minerals. Na_2S can be hydrolyzed and then dissociated in solution:

$$Na_2S + H_2O = H_2S + 2NaOH \quad (2.47)$$
$$S^{2-} + H^+ = HS^-, \qquad K_1^H = [HS^-]/[S^{2-}][H^+] = 10^{13.9} \quad (2.48)$$
$$HS^- + H^+ = H_2S, \qquad K_2^H = [H_2S]/[HS^-][H^+] = 10^{7.02},$$
$$\beta_2^H = 10^{20.92} \quad (2.49)$$
$$[S'] = [S^{2-}] + [HS^-] + [H_2S] \quad (2.50)$$
$$\phi_0 = [S^{2-}]/[S'] = 1/(1 + K_1^H[H^+] + \beta_2^H[H^+]^2)$$
$$= 1/(1 + 10^{13.9}[H^+] + 10^{20.92}[H^+]^2)$$
$$\phi_1 = [HS^-]/[S]' = K_1^H\phi_0[H^+] = 10^{13.9}\phi_0[H^+]$$
$$\phi_2 = [H_2S]/[S]' = \beta_2^H\phi_0[H^+] = 10^{20.92}\phi_0[H^+] \quad (2.51)$$

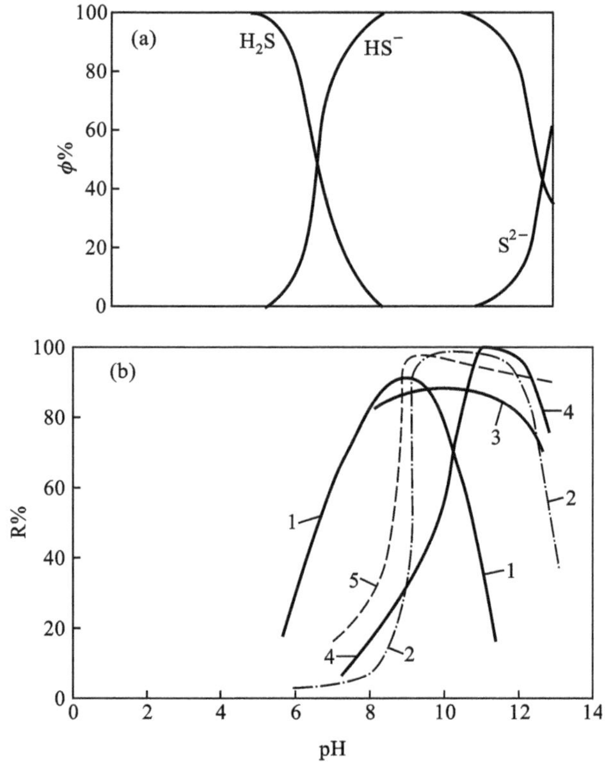

Fig. 2.1. (a) Distribution coefficients of various species in Na_2S solution as a function of pH; (b) flotation recovery of copper, lead, zinc oxide minerals as a function of pH using Na_2S as a sulfidizing agent: 1—malachite, 2—cerusite (Na_2S/DA), 3—cerusite (Na_2S/AX), 4—galmey (Na_2S/DA) and 5—smithsonite (Na_2S/DA) (Abramov, 1961; Bustamante, 1983; Marabini et al., 1984; Hu, 1986).

From Eq. (2.46), the distribution coefficients of various species containing the sulfur atom can be calculated as a function of pH, as shown in Fig. 2.1a (Bustamante et al., 1983; Wang, 1983; Marabini et al., 1984). It can be seen from this figure that H_2S dominates when pH < 7.0. whereas HS^- dominates when pH > 7.0 and S^{2-} at pH > 13.9. Flotation of copper, lead and zinc oxides using Na_2S as a sulfidizing agent is shown in Fig. 2.1b. It can be seen from a comparison of Figs. 2.1a and 2.1b that the optimum pH of flotation is located in the range where HS^- is the predominant species. It can therefore be concluded that HS^- species interact with metallic ions on the mineral surface and form a sulfidized surface film, thus activating the flotation of the oxide minerals. The sulfidization reactions of smithsonite have been experimentally shown as follows (Wang, 1983):

$$ZnCO_{3(surf)} + HS^- \rightleftharpoons ZnS_{(surf)} + HCO_3^- \tag{2.52a}$$

$$Zn(OH)_{2(surf)} + HS^- \rightleftharpoons ZnS_{(surf)} + H_2O + OH^- \tag{2.52b}$$

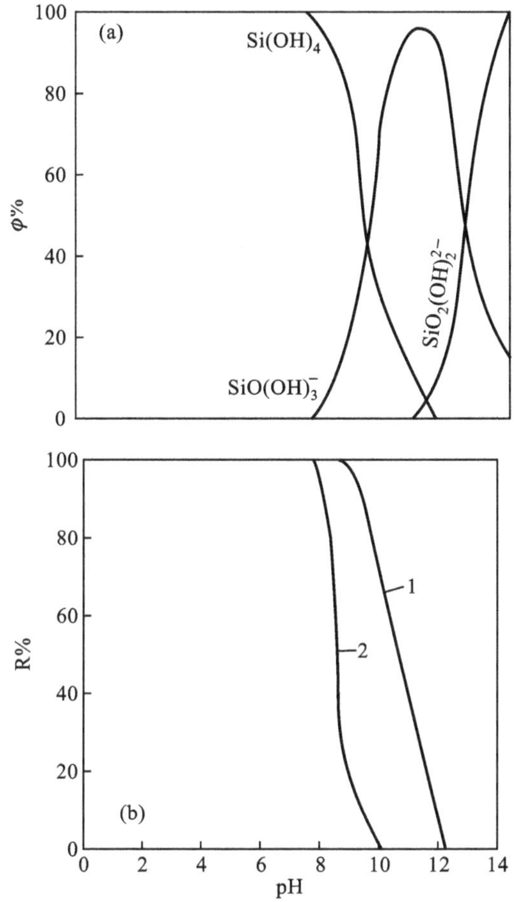

Fig. 2.2. (a) Distribution coefficients of various species of sodium silicate as a function of pH; (b) flotation recovery of the minerals as a function of pH: 1—fluorite; 2—calcite (Fuerstenau et al., 1968).

2.3.1.2. *ϕ–pH diagram and flotation effect of sodium silicates*

Sodium silicate, Na_2SiO_3, yields a number of silicate species in solution as a function of pH; the species composition in solution can vary also as a function of the ratio of the silicate in Na_2SiO_3. The solution equilibria of SiO_2 itself are as follows:

$$SiO_{2(s)(amorph)} + 2H_2O = Si(OH)_{4(aq)} \qquad K_{Si} = 10^{-2.7} \tag{2.53a}$$

$$SiO_2(OH)_2^{2-} + H^+ = SiO(OH)_3^- \qquad K_1^H = 10^{12.56} \tag{2.53b}$$

$$SiO(OH)_3^- + H^+ = Si(OH)_4 \qquad K_2^H = 10^{9.43} \tag{2.53c}$$

The ϕ–pH diagram of Na_2SiO_3 solution is shown in Fig. 2.2a (Fuerstenau et al., 1968). It can be seen that $Si(OH)_4$ is predominant when pH is less than 9.4 whereas $SiO(OH)_3^-$ is predominate when pH is greater than 9.4 and $SiO_2(OH)_2^{2-}$ is predominate when pH is above 12.6. Flotation of calcite and fluorite using sodium silicate as a depressant is shown in Fig. 2.2b. It can be seen from a comparison of Figs. 2.2a and 2.2b that the onset of pH of depression of both minerals corresponds to that of $SiO(OH)_3^-$ formation in the solution. $SiO(OH)_3^-$ can therefore be considered as the species responsible for the depression.

2.3.2. log *C*–pH diagram

From a knowledge of the original concentration (C_T) of a flotation reagent and relevant reaction constants, the absolute concentration of each species can be calculated. From Eq. (2.46),

$$[A] = \phi_0 C_T$$
$$[HA] = \phi_1 C_T$$
$$\ldots$$
$$[H_nA] = \phi_n C_T \tag{2.54}$$

2.3.2.1. log *C*–*pH diagram and flotation effect of xanthates*

Xanthate hydrolyzes in solution to form xanthic acid, HX, and then dissociates:

$$HX \rightleftharpoons H^+ + X^-, \qquad K_a = [H^+][X^-]/[HX] \tag{2.55}$$

$$\phi_0 = 1/(1 + K_a^{-1}[H^+]) \qquad [X^-] = C_T/(1 + K_a^{-1}[H^+]) \tag{2.56a}$$
$$\phi_1 = K_a^{-1}[H^+]\phi_0 \qquad [HX] = C_T[H^+]/(K_a + [H^+]) \tag{2.56b}$$

in log form:

$$\log[X^-] = \log C_T - \log(K_a + [H^+]) + \log K_a \tag{2.57a}$$
$$\log[HX] = \log C_T - pH - \log(K_a + [H^+]) \tag{2.57b}$$

where $K_a = [H^+]$, that is $pH = pK_a$

$$\log[X^-] = \log[HX] = \log C_T - \log 2 \tag{2.58}$$

The species distribution obtained using Eq. (2.58) is given in Fig. 2.3.

2.3.2.2. log *C*–*pH and flotation effect of sodium phosphate* (Na_3PO_4)

Sodium phosphate, Na_3PO_4, hydrolyzes into H_3PO_4 and then ionizes as shown below:

$$H_3PO_4 \rightleftharpoons H^+ + H_2PO_4^- \qquad K_{a1} = 1/K_3^H H = [H^+][H_2PO_4^-]/[H_3PO_4]$$
$$H_2PO_4^- \rightleftharpoons H^+ + HPO_4^{2-} \qquad K_{a2} = 1/K_2^H = [H^+][HPO_4^{2-}]/[H_2PO_4^-]$$
$$HPO_4^{2-} \rightleftharpoons H^+ + PO_4^{3-} \qquad K_{a3} = 1/K_1^H = [H^+][PO_4^{3-}]/[HPO_4^{2-}] \tag{2.59}$$

log *C*–pH diagram for this system is shown in Fig. 2.4a. At pH below 2.2, H_3PO_4 dominates whereas $H_2PO_4^-$, HPO_4^{2-} and PO_4^{3-} dominate at $2.2 < pH < 7.2$, $7.2 < pH < 12.4$ and $pH > 12.4$ respectively.

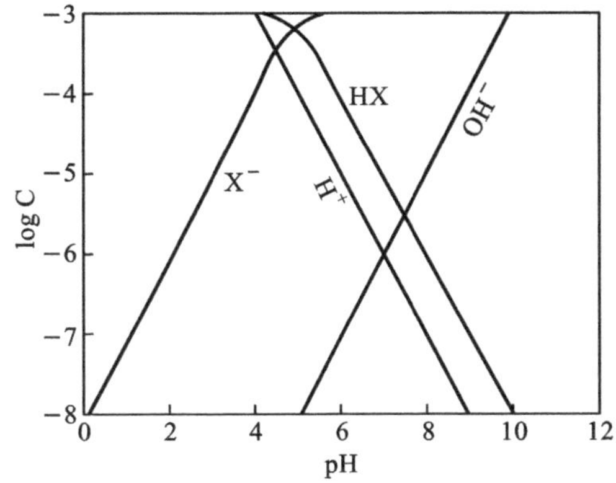

Fig. 2.3. log C–pH diagram for xanthate.

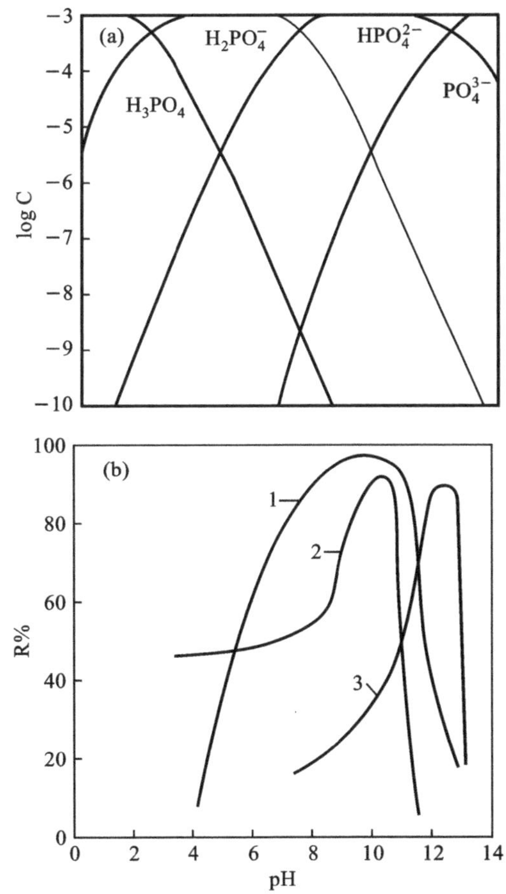

Fig. 2.4. (a) log C–pH diagram of sodium phosphate; (b) flotation recovery of the minerals as a function of pH: 1—scheelite, 2—fluorite and 3—calcite (Li and Li, 1983).

Fig. 2.4b shows the flotation of scheelite, fluorite and calcite using sodium oleate and trisodium phosphate. The flotation of fluorite and calcite can be depressed by sodium phosphate at pH $<$ 9.5 and pH $<$ 10.0, respectively, but not that of scheelite. Since the PZC of scheelite, fluorite and calcite are pH $=$ 1.5, 9.5, and 10.0, respectively (Wang, 1983), anionic species $H_2PO_4^-$, HPO_4^{2-}, and PO_4^{3-} which are predominant, will not be adsorbed on the negatively charged surface of scheelite, and thus will have no depression effect at pH $>$ 2.2; below pH 9.5, fluorite surface is positively charged and the anionic phosphate species can be adsorbed on fluorite, leading to its depression. A similar depression mechanism can be formulated for calcite at pH $<$ 10.0.

2.4. WEAK ASSOCIATION AND DISSOCIATION EQUILIBRIA OF LONG-CHAIN FLOTAGENTS

Both anionic and cationic type long-chain reagents are widely used in flotation as collectors. These include carboxylates, alkyl sulfonates, alkyl sulfates, alkyl amines and chelating agents. Most minerals, except sulfides, require long-chain collectors for their flotation. The behavior of long-chain collectors in solution is determined by the properties of the polar heads and hydrophobic tails and their resultant solvent power.

2.4.1. log C–pH diagram for oleate and amine

Weakly acidic fatty acids such as oleic acid undergo dissociation to form ions (R^-) at high pH values and neutral molecules (RH) at low pH values. In the intermediate pH region, the ions and neutral molecules can associate to form iono-molecular complexes (Kulkarni and Somasundaran, 1980). As the collector concentration is increased, micellization or precipitation of the collector will occur. In addition, collector species can undergo associative interactions to form other aggregates such as dimers (R_2^{2-}) (Somasundaran and Ananthapadmanabhan, 1979b). Since the surface activities of these species will vary from those of each other, flotation of minerals with these collectors can also be expected to be dependent upon pH and such solution conditions.

The solution equilibria of oleic acid are as follows (Somasundaran et al., 1984; Jiang et al., 2001):

$$
\begin{aligned}
&\mathrm{HOl_{(l)}} = \mathrm{HOl_{(aq)}} && S = 10^{-7.6} \\
&\mathrm{HOl_{(aq)}} = \mathrm{H^+} + \mathrm{Ol^-} && K_\mathrm{a} = [\mathrm{H^+}][\mathrm{Ol^-}]/[\mathrm{HOl_{(aq)}}] = 10^{-4.95} \\
&2\mathrm{Ol^-} = (\mathrm{Ol})_2^{2-} && K_\mathrm{d} = [(\mathrm{Ol})]_2^{2-}/[\mathrm{Ol^-}]^2 = 10^{4.0} \\
&\mathrm{HOl_{(aq)}} + \mathrm{Ol^-} = (\mathrm{Ol})_2\mathrm{H^-} && K_\mathrm{im} = [(\mathrm{Ol})_2\mathrm{H^-}]/[\mathrm{HOl_{(aq)}}][\mathrm{Ol^-}] = 10^{4.7} \qquad (2.60)
\end{aligned}
$$

$$
\text{MBE:} \quad C_\mathrm{T} = [\mathrm{HOl_{(aq)}}] + [\mathrm{Ol^-}] + 2[(\mathrm{Ol})_2^{2-}] + 2[(\mathrm{Ol})_2\mathrm{H^-}] \qquad (2.61)
$$

K_a, K_d and K_im being constants for association, dissociation and ionomolecular complexation respectively.

Substituting into Eq. (2.60) and letting $K_H = K_a/[H^+]$

$$2(K_{im}K_H + K_dK_2^H)[HOl_{(aq)}]^2 + (1 + K_H)[HOl_{(aq)}]^2 - C_T = 0 \tag{2.62}$$

The critical pH value (pH_0) for $HOl_{(l)}$ and $HOl_{(aq)}$ equilibrium is governed by the following equation:

$$2(K_{im}K_a/[H^+] + K_dK_a^2/[H^+]^2)S^2 + (1 + K_a/[H^+])S - C_T = 0$$
$$[H^+]^2 - 2.83 \times 10^{-13}[H^+]/(C_T - [S]) - 2 \times 10^{-21.1}/(C_T - S) = 0 \tag{2.63}$$

When $pH > pH_0$, from Eq. (2.62)

$$[HOl_{(aq)}] = \Big\{-(1 + K_H) + \big[(1 + K_H)^2 + 8(K_{im}K_H + K_dK_2^H)C_T\big]^{1/2}\Big\}/4(K_{im}K_H + K_dK_H^2) \tag{2.64}$$

Let

$$Z = K_{im}K_H + K_dK_H^2$$
$$W = -(1 + K_H) + [(1 + K_H)^2 + 8ZC_T]^{0.5}$$

Thus

$$[HOl_{(aq)}] = W/4Z$$
$$[Ol^-] = K_H[HOl_{(aq)}] = K_HW/4Z$$
$$[(Ol)_2^{2-}] = K_dK_2^H[HOl_{(aq)}]^2 = K_dK_2^HW^2/16Z^2$$
$$[(Ol)_2H^-] = K_{im}K^H[HOl_{(aq)}]^2 = K_dK_HW^2/16Z^2 \tag{2.65}$$

When $pH \leqslant pH_1$, from Eqs. (2.60),

$$\log[HOl_{(aq)}] = \log S$$
$$\log[Ol^-] = \log K_a + \log S + pH$$
$$\log[(Ol)_2^{2-}] = \log K_d + 2(\log K_a + \log S + pH)$$
$$\log[H(Ol)_2^-] = \log K_{im} + \log K_a + 2\log S + pH \tag{2.66}$$

Species distribution of oleic acid is shown in Fig. 2.5 as a function of pH (Somasundaran and Ananthapadmanabhan, 1979b). The features to be noted here are: (a) precipitation pH of oleic acid at the given concentration is 7.45; (b) activities of oleate monomer and dimer remain almost constant above the precipitation pH and decrease sharply below it; (c) the activity of the acid-soap (R_2H^-) exhibits a maximum in the neutral pH range. Based on charge and molecular size considerations, one can expect the surface activities of different oleate species to be markedly different. The estimation of the surface activities of various species suggests that the surface activity of the acid-soap is five orders of magnitude higher than that of the neutral molecule and about seven orders of magnitude higher than that of the oleate monomer. Experimentally, the pH dependence of the surface activity of oleate solutions can be estimated from surface tension measurements. The surface tension

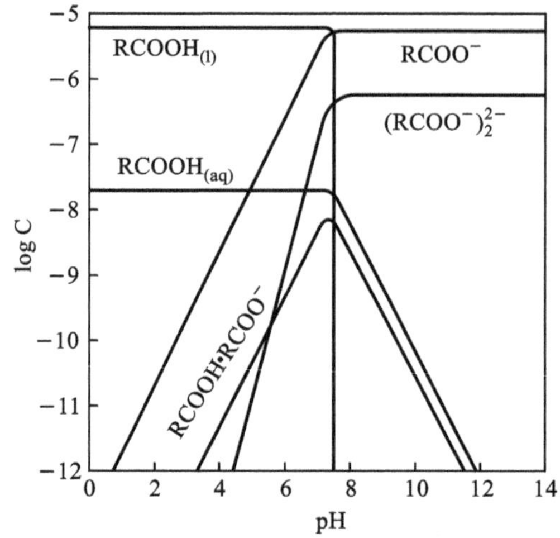

Fig. 2.5. Species distribution diagram of oleic acid (Somasundaran et al., 1984).

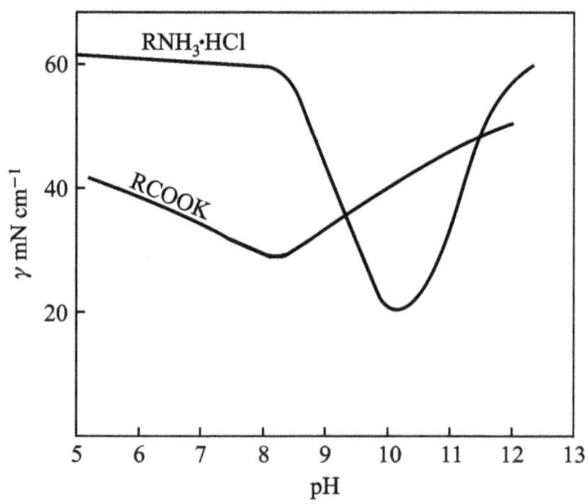

Fig. 2.6. Surface tension of potassium oleate solution and dodecylamine as a function of pH (Somasundaran and Ananthapadmanabhan, 1979b).

data given in Fig. 2.6 show oleate solutions to exhibit a minimum in surface tension in the neutral pH range where the acid-soap predominates (Somasundaran and Ananthapadmanabhan, 1979b).

The solution chemistry of cationic collectors such as dodecylamine is similar to that of oleate (Somasundaran and Ananthapadmanabhan, 1979a; Pugh and Stenius, 1985; Mukerjee, 1958; Hu and Wang, 1993). Various solution equilibria of dodecylamine are as follows:

$$RNH_{2(s)} = RNH_{2(aq)} \qquad S = 10^{-4.69}$$

$$RNH_3^+ = RNH_{2(aq)} + H^+ \qquad K_a = [H^+][RNH_{2(aq)}]/[RNH_3^+] = 10^{-10.63}$$

$$2RNH_3^+ = (RNH_3)_2^{2+} \qquad K_d = [(RNH_3)_2^{2+}]/[RNH_3^+]^2 = 10^{2.08}$$
$$RNH_3^+ = RNH_{2(aq)} = [RNH_3^+ \cdot RNH_{2(aq)}]$$

$$K_{im} = [RNH_3^+ \cdot RNH_{2(aq)}]/[RNH_3^+][RNH_2] = 10^{3.12} \tag{2.67}$$

$$\text{MBE:} \qquad C_T = [RNH_{2(aq)}] + [RNH_3^+] + 2[(RNH_3)_2^{2+}] + 2[RNH_3^+ \cdot RNH_{2(aq)}] \tag{2.68}$$

Substitution of various constants into Eq. (2.68) and $K_B = [H^+]/K_a$, yields

$$2(K_d K_B^2 + K_{im} K_B)[RNH_{2(aq)}]^2 + (1 + K_B)[RNH_{2(aq)}] - C_T = 0 \tag{2.69}$$

When $[RNH_{2(aq)}] = S$, the critical pH for amine precipitation is determined by the following equation:

$$2(K_d[H^+]^2/K_a^2 + K_{im}[H^+]/K_a)S^2 + (1 + [H^+]/K_a)S - C_T = 0$$
$$[H^+]^2 + 5.0 \times 10^{-9}[H^+] - (C_T - S)/1.82 \times 10^{14} = 0 \tag{2.70}$$

Precipitation pH_s calculated for different concentrations C_T of dodecyl amine are given in Table 2.6.

When $pH < pH_s$, from Eq. (2.69)

$$[RNH_{2(aq)}] = \{-(1 + K_B) + [(1 + K_B)^2 + 8(K_d K_B^2 + K_{im} K_B)C_T]^{1/2}\}/4(K_d K_B^2 + K_{im} K_B) \tag{2.71}$$

Let

$$X = K_d K_B^2 + K_{im} K_B$$
$$Y = -(1 + K_B) + [(1 + K_B)^2 + 8XC_T]^{1/2}$$

Then

$$[RNH_{2(aq)}] = Y/4X$$
$$[RNH_3^+] = K_B[RNH_{2(aq)}] = K_B Y/4X$$
$$[(RNH_3)_2^{2+}] = K_d[RNH_3^+] = K_d K_B^2 Y^2/16X^2$$
$$[RNH_3^+ \cdot RNH_2] = K_{im}[RNH_3^+][RNH_{2(aq)}] = K_{im} K_B Y^2/16X^2 \tag{2.72}$$

When $pH \geqslant pH_s$, from Eq. (2.67)

$$[RNH_{2(aq)}] = S = 10^{-4.69}\ \text{mol/l}$$
$$\log[RNH^+_{3(aq)}] = \log[H^+][RNH_{2(aq)}]/K_a = pK_a - pH + \log S$$
$$\log[(RNH_3^+)_2^{2+}] = \log K_d + 2pK_a - 2pH + 2\log S$$
$$\log[RNH_3^+ \cdot RNH_{2(aq)}] = \log K_{im} + pK_a - pH + 2\log S \tag{2.73}$$

When $C_T \leqslant 2.0 \times 10^{-5}$ mol/l, amine does not precipitate. When $C_T > 2 \times 10^{-5}$ mol/l, pH for amine precipitation will change with C_T. When $pH \geqslant pH_s$ concentration of amine species can be calculated using Eq. (2.73); and for $pH < pH_s$, the concentration can be calculated using Eq. (2.72). The species distribution of dodecylamine at

Table 2.6
Critical pH_S values for precipitation of dodecylamine at different concentrations

C_T mol/l	10^{-2}	10^{-3}	5×10^{-4}	10^{-4}	8×10^{-5}	5×10^{-5}	2×10^{-5}	10^{-5}
pH_S	8.04	9.04	9.32	10.06	10.27	10.48	–	–

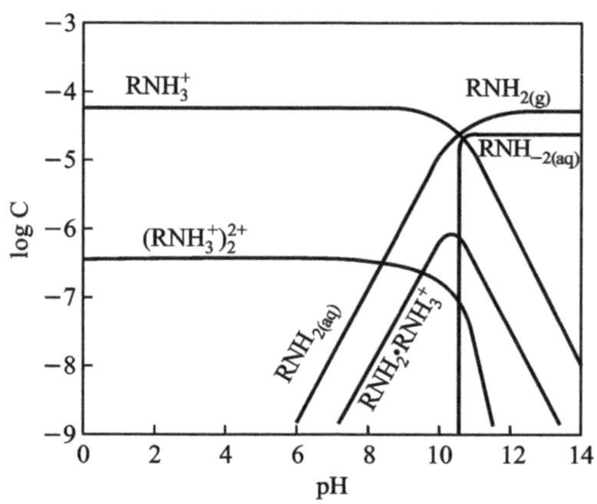

Fig. 2.7. Species distribution diagram of dodecyl amine as a function of pH (total conc. = 5×10^{-5} mol/l).

$C_T = 1.0 \times 10^{-5}$ mol/l is shown in Fig. 2.7. In this case the neutral molecule (RNH_2) precipitates in the high pH range and the ionic forms RNH_3^+ and $(RNH_3)_2^{2+}$ dominate in the acidic pH range. Again, the ion-molecular complex exhibits a maximum at certain pH values. In this case also, the pH dependence of surface tension of amine solution and the flotation behavior of silica using amine have been correlated with the activities of the species.

Similar estimates of species distribution have been made also for sodium dodecyl sulfonate (SDS) (Somasundaran and Ananthapadmanabhan, 1979b), as shown in Fig. 2.8.

In contrast to the acid-soap molecule, the ionic dimer (R_2^{2-}) will have the two ionic heads as far apart as possible with maximum contact between the hydrocarbon chains. From the flotation point of view, adsorption of the dimer may not be desirable, as it may not make the surface sufficiently hydrophobic.

Interestingly, the flotation often decreases at high concentration of the collector. At much higher concentrations, collectors normally associate to form micelles in solutions. The activity of collector monomer does not increase appreciably above the critical micelle concentration (CMC); in any case, the concentration of interest in the case of flotation is usually in the pre-micellar region. Adsorption in the micellar region is of interest in other areas such as enhanced oil recovery. Detailed discussions are given in the next chapter.

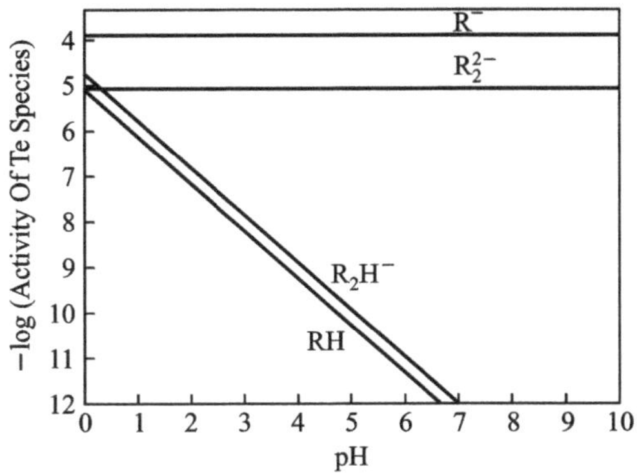

Fig. 2.8. Species distribution diagram of sodium dodecyl sulfonate as a function of pH (total conc. = 1×10^{-5} mol/l) (Somasundaran and Ananthapadmanabhan, 1979b).

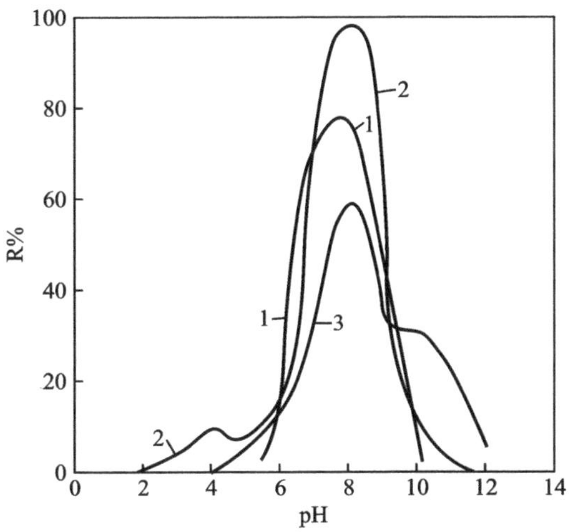

Fig. 2.9. Flotation recovery of the minerals using oleate as a function of pH: 1—fluorite, 2—spodumene, 3—augite (Fuerstenau, 1977, 1980; Pugh and Stenius, 1985).

2.4.2. Role of associated species (ion–molecule complexes) in flotation

As discussed above, long-chain collectors can form ion–molecule complexes which exhibit relatively high surface activity in a certain pH range. Similar to lowering of surface tension, oleate flotation of hematite and a number of other minerals exhibits a maximum in the same pH range (pH = 7–9), as shown in Figs. 2.9 and 2.10 (Fuerstenau, 1977, 1980; Pugh and Stenius, 1985). This clearly shows that oleate solutions impart maximum surface activity to minerals in the pH region where the oleate is most surface active.

In the case of flotation of hematite using oleate, the pH value at which the flotation exhibits the maximum depends upon the concentration of oleate (see Fig. 2.11) pH of

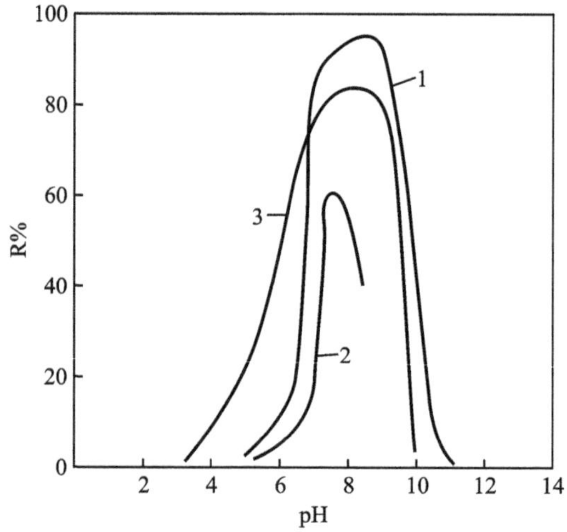

Fig. 2.10. Flotation recovery of additional minerals using oleate as a function of pH: 1—scheelite, 2—taconite and 3—wolframite (Somasundaran and Ananthapadmanabhan, 1979b; Li and Li, 1983; Hu, 1984).

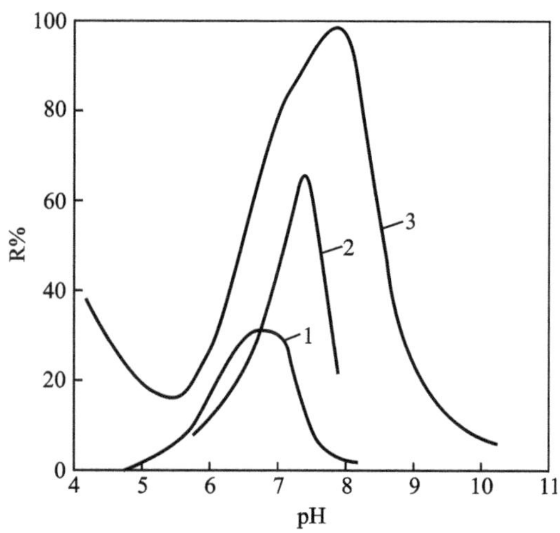

Fig. 2.11. Flotation recovery of hematite using oleate: (1) 7.5×10^{-6} mol/l (2) 1.5×10^{-5} mol/l and (3) 3.0×10^{-5} mol/l (Somasundaran and Ananthapadmanabhan, 1979b).

maximum flotation recovery. The pH of maximum ion molecules is plotted in Fig. 2.12 at three different oleate concentrations. A linear relationship between the concentration of ion–molecule complex and floatability of minerals is apparent.

Similar to the above, in the case of flotation of flotation using dodecylamine (Fig. 2.13), the pH at which maximum flotation corresponds to that of the formation of maximum

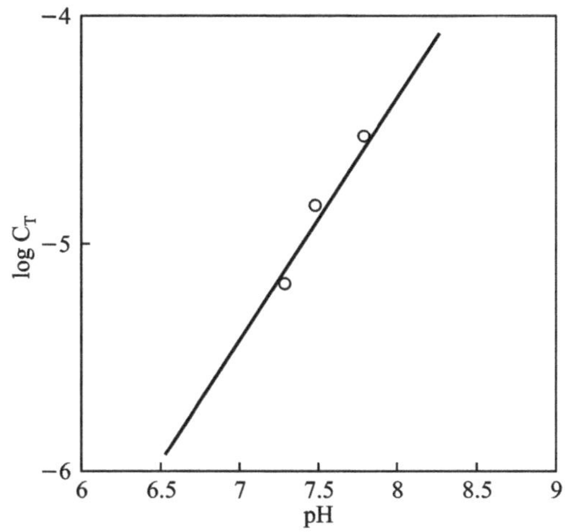

Fig. 2.12. pH values of maximum ion-molecular complex formation and total concentration of correlating oleate for flotation.

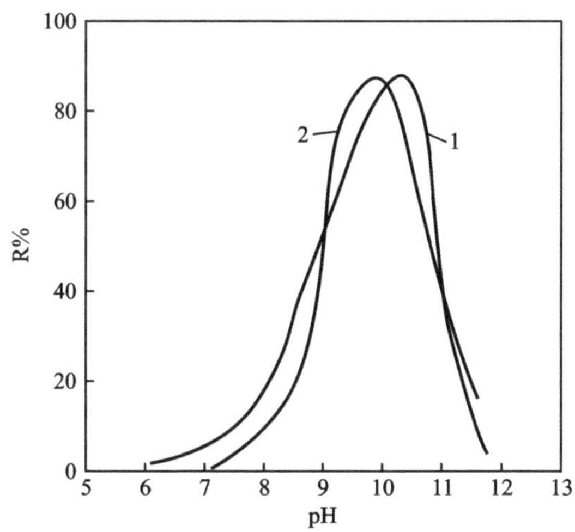

Fig. 2.13. Flotation recovery of quartz using dodecyl amine: (1) 1.0×10^{-5} mol/l, (2) 5.0×10^{-5} mol/l.

amounts of ion–molecule complex $RNH_2 \cdot RNH_3^+$. The pH of maximum flotation in the case also depends on the concentration of dodecylamine, with a relationship similar to that between pH of *ion–molecule* complex formation and the amine concentration.

From the above discussion, it is clear that long-chain flotation reagents characterized by weak association can form strong surface active ion–molecule complexes which are evidently the most efficient species for the flotation of many minerals.

2.5. STRONG ASSOCIATION AND MICELLIZATION EQUILIBRIA OF LONG-CHAIN SURFACTANTS

2.5.1. Thermodynamics of micellization (Arai et al., 1971; Rosen, 1978)

Solution properties of long-chain collectors are closely related to the bulk properties of the solvent medium itself. It has been shown experimentally that physico-chemical properties of aqueous solution of surfactants (long-chain flotagents) depend upon their concentration in solution, as shown in Fig. 2.14 (Preston, 1948). It can be seen from this figure that a number of physico-chemical properties of surfactants show a marked change in shape within a narrow range concentration. This change has been attributed to association of the surfactants into aggregates called micelles and the concentration at which micellization occurs has been called the critical micelle concentration (CMC).

Based on the vast amount of research that has been done on micellization, an understanding of the process of micellization as well as the structure of the micelles has slowly emerged. When dissolved in a solvent, flotation reagents that contain a hydrophobic group and at least one hydrophilic group in the same molecule disrupt the interactions between the solvent molecules distorting the structure of the solvent and thus increase the free energy of the system. The distortion of the solvent structure can be decreased by the aggrega-

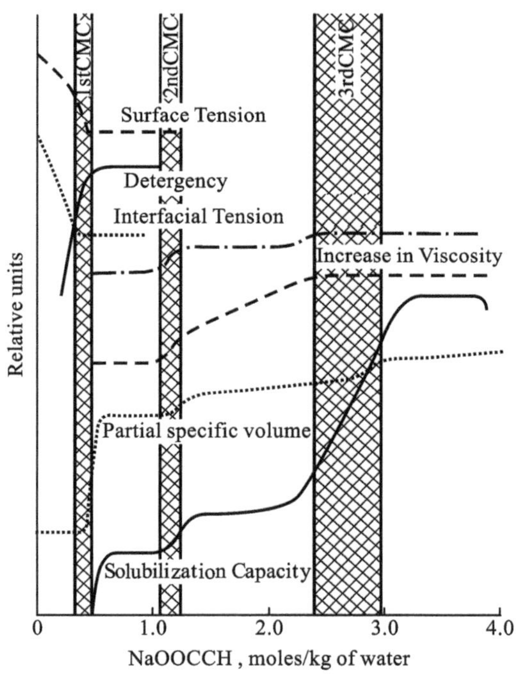

Fig. 2.14. Variation of physical properties of solutions with concentration of sodium octanoate (Preston, 1948).

tion of the surface-active species into clusters (micelles) with their hydrophobic groups directed toward the interior of the micelle and the hydrophilic groups towards the solvent. Micellization can thus be considered to be a mechanism alternative to adsorption at the interfaces for removing hydrophobic groups from contact with the solvent. There are two types of thermodynamic models which describe the formation of micelles (Rosen, 1978; Chen and Dai, 1985).

2.5.1.1. *Phase separation model*

This model considers micellization as a phase separation process which begins at CMC (saturation concentration of surfactant species in non-micellar form) with micelles as the new phase formed.

For an ionic surfactant, the equilibrium of micelle formation is:

$$jC^+ + jA^- = M, \tag{2.74}$$

where C^+ and A^- respectively represent cations and anions of the surfactant which form micelles, M, with an aggregation number, j. The free energy change of micelle formation (per mole of monomer) is given by:

$$\Delta G_p^0 = -(RT/j)\ln(a_M/a_+a_-) \tag{2.75}$$

When micelles are present as a new phase, $a_M = 1$; $a_+ = a_- \approx$ CMC since the concentration of micelles at CMC is quite small. Then

$$\Delta G_p^0 = 2RT \ln \mathrm{CMC} \tag{2.76}$$

For a nonionic surfactant

$$\Delta G_p^0 = RT \ln \mathrm{CMC} \tag{2.77}$$

From Gibbs–Helmholtz equation, enthalpy change of micellization is:

$$\Delta H_p^0 = -T(\partial \Delta G_p^0/\partial T)_p = -2RT^2(\partial \ln \mathrm{CMC}/\partial T) \tag{2.78}$$

Corresponding entropy change is:

$$\Delta S_p^0 = (\Delta H_p^0 - \Delta G_p^0)/T = -2R \ln \mathrm{CMC} - 2RT(\partial \ln \mathrm{CMC}/\partial T)_p \tag{2.79}$$

2.5.1.2. *Mass action model*

In the treatment by the mass action model, micellization is considered as an association-dissociation equilibrium of individual molecules or ions with micelles in the concentration range above CMC.

For aggregation equilibrium of ionic surfactants

$$jC^+ + (j - Z)A^- = M^{Z+} \tag{2.80}$$

where M^{Z+} is a micellar aggregate with j cations and $j - Z$ counterions. The equilibrium constant in dilute solution is:

$$K_m = [M^{Z+}]/[C^+]^j[A^-]^{j-Z} \tag{2.81}$$

$$\Delta G_m^0 = (-RT/j)\ln([M^{Z+}]/[C^+]^j[A^-]^{j-Z}) \quad (2.82)$$

In dilute solutions, $(1/j)\ln[M^{Z+}]$ can be neglected; $[C^+] = [A^-] =$ CMC. The free energy change of micellization is then given by

$$\Delta G_m^0 = (2 - Z/j)RT \ln \mathrm{CMC} \quad (2.83)$$

Two limiting conditions of counterions binding to micelle is given by $Z = 0$ and $Z = j$

(a) When $Z = 0$, all counterions are bound to the micelles.

$$\Delta G_m^0 = 2RT \ln \mathrm{CMC} \quad (2.84)$$

(b) When $Z = j$, no counterions are bound to the micelles and all are in the diffuse layer.

$$\Delta G_m^0 = RT \ln \mathrm{CMC} \quad (2.85)$$

For a nonionic surfactant

$$\Delta G_m^0 = RT \ln \mathrm{CMC} \quad (2.86)$$

According to the mass action model, the enthalpy and entropy changes are respectively given by:

$$\Delta H_m^0 = -(2 - Z/j)RT^2(\delta \ln \mathrm{CMC}/\delta T)_p$$
$$\Delta S_m^0 = -(2 - Z/j)R[\ln \mathrm{CMC} + T(\delta \ln \mathrm{CMC}/\delta T)_p] \quad (2.87)$$

In the case of an equilibrium of individual molecules or ions with micelles, the chemical potential of the surfactant species in the bulk μ_s must be equal to its chemical potential in the micelle μ_m.

$$\mu_s = \mu_m \quad (2.88)$$

Using statistical thermodynamics, the following equation can be derived:

$$\ln \mathrm{CMC} = \ln(1000/N_0 V) - (\omega/kT) - 1, \quad (2.89)$$

where k is the Boltzman constant, N_0 is Avogadro constant, V is free volume of molecules and ω represents potential difference between the micelle state and the disperse state. Including interaction energy of molecules $n_i\phi$ and electrical energy E_{el}, where n is carbon number in the straight hydrocarbon chain and ϕ is interaction energy per each $-CH_2-$ group, for ionic surfactants:

$$\ln \mathrm{CMC} = \ln 1000/N_0 V - 1 - n_i\phi/kT + E_{el}/kT \quad (2.90)$$

E_{el} can be calculated using the electrical double layer theory and Possion equation.

$$E_{el} = K_0 e\psi_0 = K_0 kT(\ln 2000\pi\sigma^2/\varepsilon N_0 kT - \ln C_i), \quad (2.91)$$

where σ is surface charge density, ε is dielectric constant, C_i is molar concentration of counterions, K_0 is the charge fraction of the individual ions consisting of micelles, which is determined by the association of counterions.

Substituting Eq. (2.91) into (2.90)

$$\ln \mathrm{CMC} = A_1 - n_i\phi/kT + K_0(\ln 2000\pi\sigma^2/\varepsilon N_0 kT - \ln C_i), \tag{2.92}$$

where

$$A_1 = \ln 1000/N_0 V - 1,$$

or

$$\ln \mathrm{CMC} = -n_i\phi/kT - K_0 \ln C_i + A' \tag{2.93}$$

$$A' = A_1 + K_0 \ln(2000\pi\sigma^2/\varepsilon N_0 kT)$$

For nonionic surfactant, $E_{el} = 0$

$$\ln \mathrm{CMC} = A_1 - n_i\phi/kT \tag{2.94}$$

The relationships between CMC and various thermodynamic functions discussed above are also relevant to the adsorption of surfactants at solid/liquid interfaces and will be discussed later.

2.5.2. Structure and aggregation number of micelles

The structure of micelles in aqueous medium, at concentrations not too far above the CMC and in the absence of additives that are solubilized by the micelle, can be considered to be roughly spherical with a core region consisting of the hydrophobic groups of the surfactant, and the radius equal to approximately the length of a fully extended hydrophobic group surrounded by an outer region consisting of the hydrated hydrophilic groups and bound water. In the case of ionic micelles, the outer region consists of the ionic head groups surrounded by counterions in the electrical double layer around the head groups.

Changes in temperature, concentration of the surfactant, additives in the liquid phase, and functional groups in the surfactant can all cause changes in the size, shape and aggregation number of the micelle. The shape can vary from spherical through rod or disklike to lamellar (Shah et al., 1977), as shown in Fig. 2.15.

In concentrated solutions, at ten times the CMC or more, micelles are generally nonspherical. At least in some cases, the surfactant molecules are believed to form extended parallel sheets of two molecules thick (lamellar micelles) with the individual molecules oriented perpendicular to the plane of the sheet. In aqueous solutions, the hydrophilic heads of the surfactant molecules form the two parallel surfaces of the sheets and the hydrophobic tails comprise the interior region.

Aggregation number of surfactant micelles is an important characteristic parameter which also describes the size of the micelles formed. Generally, in aqueous medium, greater the "dissimilarity" between the surfactant and the solvent, greater is the aggregation number. Thus the micellar aggregation number appears to increase with an increase in the hydrophobic character of the surfactant owing to any increase in the length of the hydrophobic group or a decrease in the number of hydrophilic groups such as oxyethylene

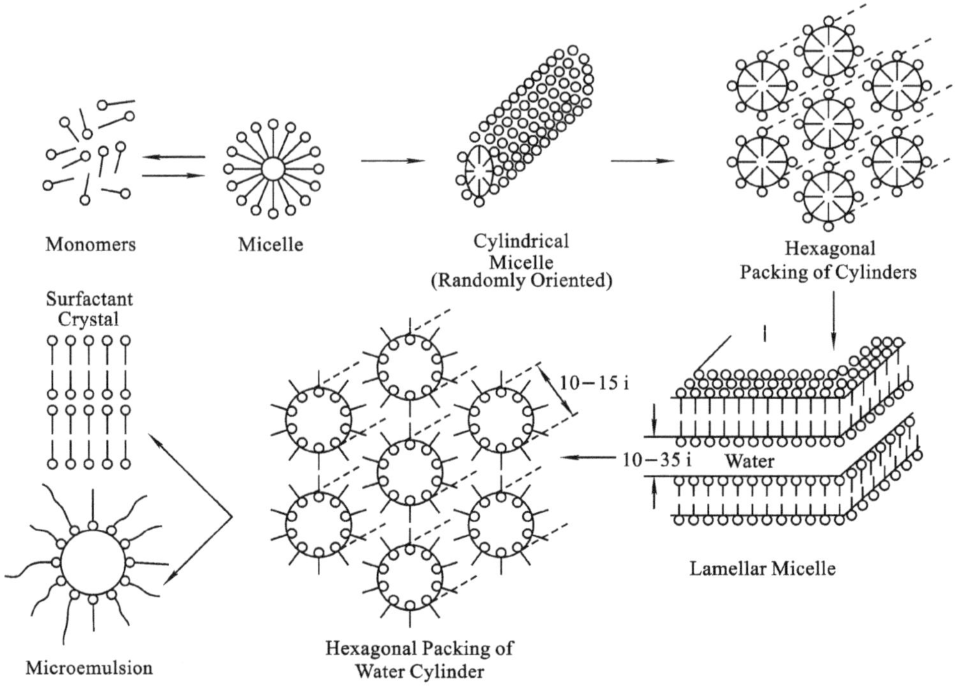

Fig. 2.15. Different structures of micelles and mesophases (Shah et al., 1977).

units in the polyoxyethylenated nonionics. Micellar aggregation numbers are generally determined by light scattering and ultracentrifuge techniques and, recently, fluorescence spectroscopy. The CMC and the aggregation number of different types of common surfactants are listed in Tables 2.7a and 2.7b (Rosen, 1978, 2004).

2.5.3. Calculations of CMC and flotation

Equations for estimating the CMC of various surfactants are classified into two types: theoratical and empirical.

2.5.3.1. *Theoretical*

Equations relating CMC to the free energy change, ΔG^0_{m}, of the aggregation of the individual surfactant molecules that form micelles can be derived on the basis of the theoretical considerations of Shinoda (1963). When no salt is added to the surfactant solution, C_i = CMC, and Eq. (2.93) becomes:

$$\log \mathrm{CMC} = A' - n\phi/2.303kT(1 + K_0) \tag{2.95}$$

For a nonionic surfactant

$$\log \mathrm{CMC} = A_1 - n\phi/2.303kT \tag{2.96}$$

Table 2.7a
Critical micelle concentrations of some surfactants in aqueous media

Compound	Solvent	T (°C)	CMC (M)	Reference
Anionics				
$C_8H_{17}SO_3^-Na^+$	H_2O	40	1.6×10^{-1}	Klevens (1948)
$C_8H_{17}SO_4^-Na^+$	H_2O	40	1.4×10^{-1}	Evans (1956)
$C_{10}H_{21}SO_3^-Na^+$	H_2O	40	4.1×10^{-2}	Klevens (1948)
$C_{10}H_{21}SO_4^-Na^+$	H_2O	40	3.3×10^{-2}	Evans (1956)
$C_{11}H_{23}SO_4^-Na^+$	H_2O	21	1.6×10^{-2}	Huisman (1961)
$C_{12}H_{25}SO_3^-Na^+$	H_2O	40	9.7×10^{-3}	Bujake (1965)
$C_{12}H_{25}SO_4^-Na^+$	H_2O	40	8.6×10^{-3}	Flockhart (1961)
$C_{13}H_{27}SO_4^-Na^+$	H_2O	40	4.3×10^{-3}	Gotte (1969)
$C_{14}H_{29}SO_3^-Na^+$	H_2O	40	2.5×10^{-3}	Klevens (1948)
$C_{14}H_{29}SO_4^-Na^+$	H_2O	40	2.2×10^{-3}	Flockhart (1961)
$C_{15}H_{31}SO_4^-Na^+$	H_2O	40	1.2×10^{-3}	Gotte (1969)
$C_{16}H_{33}SO_3^-Na^+$	H_2O	50	7×10^{-4}	Klevens (1948)
$C_{16}H_{33}SO_4^-Na^+$	H_2O	40	5.8×10^{-4}	Evans (1956)
$C_{18}H_{37}SO_4^-Na^+$	H_2O	50	2.3×10^{-4}	Gotte (1960)
$C_{12}H_{25}SO_4^-Na^+$	H_2O	25	8.2×10^{-3}	Elworthy (1966)
$C_{12}H_{25}SO_4^-Li^+$	H_2O	25	8.9×10^{-3}	Mysels (1959)
$C_{12}H_{25}SO_4^-K^+$	H_2O	40	7.8×10^{-3}	Meguro (1956)
$(C_{12}H_{25}SO_4^-)^2Ca^{2+}$	H_2O	70	3.4×10^{-3}	Corkill (1962)
$C_{12}H_{25}SO_4^-N(CH_3)_4^+$	H_2O	25	5.5×10^{-3}	Mysels (1959)
$C_{12}H_{25}SO_4^-N(C_2H_5)_4^+$	H_2O	30	4.5×10^{-3}	Meguro (1956)
$C_{12}H_{25}SO_4^-N(C_3H_7)_4^+$	H_2O	25	2.2×10^{-3}	Mukerjee (1967)
$C_{12}H_{25}SO_4^-N(C_4H_9)_4^+$	H_2O	30	1.3×10^{-3}	Meguro (1959)
$C_{12}H_{25}SO_4^-Na^+$	3 M urea	25	9.0×10^{-3}	Schick (1964)
$C_{12}H_{25}SO_4^-N(CH_3)_4^+$	3 M urea	25	6.9×10^{-3}	Schick (1964)
$C_{12}H_{25}SO_4^-Na^+$	0.01 M NaCl	21	5.6×10^{-3}	Huisman (1964)
$C_{12}H_{25}SO_4^-Na^+$	0.03 M NaCl	21	3.2×10^{-3}	Huisman (1964)
$C_{12}H_{25}SO_4^-Na^+$	0.1 M NaCl	21	1.5×10^{-3}	Huisman (1964)
$C_{16}H_{33}SO_4^-Na^+$	H_2O	40	5.8×10^{-4}	Evans (1956)
$C_{12}H_{25}CH(SO_4^-Na^+)C_3H_7$	H_2O	40	1.7×10^{-3}	Evans (1956)
$C_{10}H_{21}CH(SO_4^-Na^+)C_5H_{11}$	H_2O	40	2.4×10^{-3}	Evans (1956)
$C_8H_{17}CH(SO_4^-Na^+)C_7H_{15}$	H_2O	40	4.3×10^{-3}	Evans (1956)
$C_{13}H_{27}CH(CH_3)CH_2SO_4^-Na^+$	H_2O	40	8.0×10^{-4}	Gotte (1969)
$C_{12}H_{25}CH(C_2H_5)CH_2SO_4^-Na^+$	H_2O	40	9.0×10^{-4}	Gotte (1969)
$C_{11}H_{23}CH(C_3H_7)CH_2SO_4^-Na^+$	H_2O	40	1.1×10^{-3}	Gotte (1969)
$C_{10}H_{21}CH(C_4H_9)CH_2SO_4^-Na^+$	H_2O	40	1.5×10^{-3}	Gotte (1969)
$C_9H_{19}CH(C_5H_{11})CH_2SO_4^-Na^+$	H_2O	40	2×10^{-3}	Gotte (1969)
$C_8H_{17}CH(C_6H_{13})CH_2SO_4^-Na^+$	H_2O	40	2.3×10^{-3}	Gotte (1969)
$C_7H_{15}CH(C_7H_{15})CH_2SO_4^-Na^+$	H_2O	40	3×10^{-3}	Gotte (1969)
$C_{12}H_{25}OC_2H_4SO_4^-Na^+$	H_2O	50	4.8×10^{-3}	Gotte (1960)
$C_{12}H_{25}(OC_2H_4)_2SO_4^-Na^+$	H_2O	50	3.0×10^{-3}	Gotte (1960)
$C_{12}H_{25}(OC_2H_4)_3SO_3^-Na^+$	H_2O	50	2.0×10^{-3}	Gotte (1960)
$C_{12}H_{25}(OC_2H_4)_4SO_4^-Na^+$	H_2O	50	1.3×10^{-3}	Gotte (1960)
$C_8H_{17}OOC(CH_2)_2SO_3^-Na^+$	H_2O	30	4.6×10^{-3}	Hikota (1970)
$C_{10}H_{21}OOC(CH_2)_2SO_3^-Na^+$	H_2O	30	1.1×10^{-3}	Hikota (1970)
$C_{12}H_{25}OOC(CH_2)_2SO_3^-Na^+$	H_2O	30	2.2×10^{-3}	Hikota (1970)

Table 2.7a
(*Continued*)

Compound	Solvent	T (°C)	CMC (M)	Reference
$C_{14}H_{29}OOC(CH_2)_2SO_3^-Na^+$	H_2O	40	9×10^{-4}	Hikota (1970)
$C_4H_9OOCCH_2CH(SO_3^-Na^+)COOC_4H_9$	H_2O	25	2.0×10^{-1}	Williams (1957)
$C_5H_{11}OOCCH_2CH(SO_3^-Na^+)COOC_5H_{11}$	H_2O	25	5.3×10^{-2}	Williams (1957)
$C_6H_{13}OOCCH_2CH(SO_3^-Na+)COOC_6H_{13}$	H_2O	25	1.2×10^{-2}	Williams (1957)
$C_4H_9CH(C_2H_5)CH_2OOCCH_2-$ $CH(SO_3^-Na^+)COOCCH_2CH(C_2H_5)C_4H_9$	H_2O	25	2.5×10^{-3}	Williams (1957)
$C_8H_{17}OOCCH_2CH(SO_3^-Na^+)COOC_8H_{17}$	H_2O	25	6.8×10^{-4}	Williams (1957)
p-n-$C_8H_{17}C_6H_4SO_3^-Na^+$	H_2O	35	1.5×10^{-2}	Gershman (1957)
p-n-$C_{10}H_{21}C_6H_4SO_3^-Na^+$	H_2O	50	3.1×10^{-3}	Gershman (1957)
p-n-$C_{12}H_{25}C_6H_4SO_3^-Na^+$	H_2O	60	1.2×10^{-3}	Gershman (1957)
Fluorinated anionics				
$C_7H_{15}COO^-K^+$	H_2O	25	2.9×10^{-2}	Shinoda (1964)
$C_7H_{15}\ COO^-Na^+$	H_2O	25	3.0×10^{-2}	Shinoda (1972)
$(CF_3)_2CF(CF_2)_4COO^-Na^+$	H_2O	25	3.0×10^{-2}	Shinoda (1972)
n-$C_8F_{17}\ SO_3^-Li^+$	H_2O	25	6.3×10^{-2}	Shinoda (1972)
Cationics				
$C_8H_{17}N(CH_3)_3^+Br^-$	H_2O	25	1.4×10^{-1}	Klevens (1948)
$C_{10}H_{21}N(CH_3)_3^+Br^-$	H_2O	25	6.8×10^{-2}	Klevens (1948)
$C_{10}H_{21}N(CH_3)_3^+Cl^-$	H_2O	25	6.1×10^{-2}	Hoyer (1961)
$C_{12}H_{25}N(CH_3)_3^+Br^-$	H_2O	25	1.6×10^{-2}	Klevens (1948)
$C_{12}H_{25}N(CH_3)_3^+Cl^-$	H_2O	25	2.0×10^{-2}	Osugi (1965)
$C_{12}H_{25}^+N(C_6H_5)Cl^-$	H_2O	25	1.5×10^{-2}	Ford (1968)
$C_{14}H_{29}N(CH_3)_3^+Br^-$	H_2O	30	3.5×10^{-3}	Venable (1964)
$C_{14}H_{29}N(CH_3)_3^+Cl^-$	H_2O	25	4.5×10^{-3}	Hoyer (1961)
$C_{14}H_{29}N(C_3H_7)_3^+Br^-$	H_2O	30	2.1×10^{-3}	Venable (1964)
$C_{14}H_{29}^+NC_6H_4\ Br^-$	H_2O	30	2.6×10^{-3}	Venable (1964)
$C_{16}H_{33}N(CH_3)_3^+Br^-$	H_2O	25	9.2×10^{-4}	Czerniawski (1968)
$C_{16}H_{33}N(CH_3)_3^+Cl^-$	H_2O	30	1.3×10^{-3}	Ralston (1974)
$C_{16}H_{33}^+NC_6H_4Cl^-$	H_2O	25	9.0×10^{-4}	Hartley (1938)
$C_{18}H_{33}^+NC_6H_4Cl^-$	H_2O	25	2.4×10^{-4}	Evers (1948)
$C_{12}H_{25}N(CH_3)_3^+F^-$	0.5 M NaF	31.5	8.4×10^{-3}	Anacker (1963)
$C_{12}H_{25}N(CH_3)_3^+Cl^-$	0.5 M NaCl	31.5	3.8×10^{-3}	Anacker (1963)
$C_{12}H_{25}N(CH_3)_3^+Br^-$	0.5 M NaBr	31.5	1.9×10^{-3}	Anacker (1963)
$C_{12}H_{25}N(CH_3)_3^+NO_3^+$	0.5 M $NaNO_3$	31.5	8×10^{-3}	Anacker (1963)
Anionic-cationic salts				
$C_6H_{13}\ SO_4^- \cdot {}^+N(CH_3)_3C_6H_{13}$	H_2O	25	1.1×10^{-1}	Corkill (1966)
$C_8H_{17}SO_4^- \cdot {}^+N(CH_3)_3C_8H_{17}$	H_2O	25	7.5×10^{-3}	Corkill (1965)
$C_{10}H_{21}SO_4^- \cdot {}^+N(CH_3)_3C_{10}H_{21}$	H_2O	25	4.6×10^{-4}	Corkill (1963)
$C_{12}H_{25}SO_4^- \cdot {}^+N(CH_3)_3C_{12}H_{25}$	H_2O	25	3×10^{-5}	Hoyer (1961)
Zwitterionics				
$C_8H_{17}^+N(CH_3)_2CH_2COO^-$	H_2O	27	2.5×10^{-1}	Tori (1963a)
$C_8H_{17}CH(COO^-)N^+(CH_3)_3$	H_2O	27	$9.7 \times 10^{-}$	Tori (1963a)
$C_8H_{17}CH(COO^-)N^+(CH_3)_3$	H_2O	60	8.6×10^{-2}	Tori (1963b)
$C_{10}H_{21}CH(COO^-)N^+(CH_3)_3$	H_2O	27	1.3×10^{-2}	Tori (1963b)
$C_{12}H_{25}CH(COO^-)N^+(CH_3)_3$	H_2O	27	1.3×10^{-3}	Tori (1963b)

Table 2.7a
(*Continued*)

Compound	Solvent	T (°C)	CMC (M)	Reference
Nonionics				
n-$C_4H_9(OC_2H_4)_6OH$	H_2O	20	8.0×10^{-1}	Elworthy (1964)
n-$C_4H_9(OC_2H_4)_6OH$	H_2O	40	7.1×10^{-1}	Elworthy (1964)
$(CH_3)_2CHCH_2(OC_2H_4)_6OH$	H_2O	20	9.1×10^{-1}	Elworthy (1964)
$(CH_3)_2CHCH_2(OC_2H_4)_6OH$	H_2O	40	8.5×10^{-1}	Elworthy (1964)
n-$C_6H_{13}(OC_2H_4)_6OH$	H_2O	20	7.4×10^{-2}	Elworthy (1964)
n-$C_6H_{13}(OC_2H_4)_6OH$	H_2O	40	5.2×10^{-2}	Elworthy (1964)
$(C_2H_5)_2CHCH_2(OC_2H_4)_6OH$	H_2O	20	1.0×10^{-1}	Elworthy (1964)
$(C_2H_5)_2CHCH_2(OC_2H_4)_6OH$	H_2O	40	8.7×10^{-2}	Elworthy (1964)
$(C_3H_7)_2CHCH_2(OC_2H_4)_6OH$	H_2O	20	2.3×10^{-2}	Elworthy (1964)
$(C_4H_9)_2CHCH_2(OC_2H_4)_6OH$	H_2O	20	3.1×10^{-3}	Elworthy (1964)
$(C_4H_9)_2CHCH_2(OC_2H_4)_9OH$	H_2O	20	3.2×10^{-3}	Elworthy (1964)
n-$C_{12}H_{25}(OC_2H_4)_4OH$[a]	H_2O	25	4×10^{-5}	Elworthy (1964)
n-$C_{12}H_{25}(OC_2H_4)_4OH$[a]	H_2O	55	1.7×10^{-5}	Elworthy (1964)
n-$C_{12}H_{25}(OC_2H_4)_7OH$[a]	H_2O	25	5×10^{-5}	Elworthy (1964)
n-$C_{12}H_{25}(OC_2H_4)_7OH$[a]	H_2O	55	2×10^{-5}	Schick (1965)
n-$C_{12}H_{25}(OC_2H_4)_{14}OH$[a]	H_2O	25	3.5×10^{-5}	Schick (1965)
n-$C_{12}H_{25}(OC_2H_4)_{14}OH$[a]	H_2O	55	2.5×10^{-5}	Schick (1965)
n-$C_{12}H_{25}(OC_2H_4)_{23}OH$[a]	H_2O	25	6.0×10^{-5}	Schick (1965)
n-$C_{12}H_{25}(OC_2H_4)_{23}OH$[a]	H_2O	55	3.0×10^{-5}	Schick (1965)
n-$C_{12}H_{25}(OC_2H_4)_{31}OH$[a]	H_2O	25	8.0×10^{-5}	Schick (1965)
n-$C_{12}H_{25}(OC_2H_4)_{31}OH$[a]	H_2O	55	4.0×10^{-5}	Schick (1965)
$C_{16}H_{33}(OC_2H_4)_7H$	H_2O	25	1.7×10^{-6}	Elworthy (1962)
$C_{16}H_{33}(OC_2H_4)_9H$	H_2O	25	2.1×10^{-6}	Elworthy (1962)
$C_{16}H_{33}(OC_2H_4)_{12}H$	H_2O	25	2.3×10^{-6}	Elworthy (1962)
$C_{16}H_{33}(OC_2H_4)_{15}H$	H_2O	25	3.1×10^{-6}	Elworthy (1962)
$C_{16}H_{33}(OC_2H_4)_{21}H$	H_2O	25	3.9×10^{-6}	Elworthy (1962)
p-l-$C_8H_{17}C_6H_4O(C_2H_4O)_2H$	H_2O	25	1.3×10^{-4}	Crook (1963)
p-l-$C_8H_{17}C_6H_4O(C_2H_4O)_3H$	H_2O	25	9.7×10^{-5}	Crook (1963)
p-l-$C_8H_{17}C_6H_4O(C_2H_4O)_4H$	H_2O	25	1.3×10^{-4}	Crook (1963)
p-l-$C_8H_{17}C_6H_4O(C_2H_4O)_5H$	H_2O	25	1.5×10^{-4}	Crook (1963)
p-l-$C_8H_{17}C_6H_4O(C_2H_4O)_6H$	H_2O	25	2.1×10^{-4}	Crook (1963)
p-l-$C_8H_{17}C_6H_4O(C_2H_4O)_7H$	H_2O	25	2.5×10^{-4}	Crook (1963)
p-l-$C_8H_{17}C_6H_4O(C_2H_4O)_8H$	H_2O	25	2.8×10^{-4}	Crook (1964)
p-l-$C_8H_{17}C_6H_4O(C_2H_4O)_9H$	H_2O	25	3.0×10^{-4}	Crook (1964)
p-l-$C_8H_{17}C_6H_4O(C_2H_4O)_{10}H$	H_2O	25	3.3×10^{-4}	Crook (1964)
$C_9H_{19}C_6H_4(OC_2H_4)_{10}OH$[a,b]	H_2O	25	7.5×10^{-5}	Schick (1965)
$C_9H_{19}C_6H_4(OC_2H_4)_{10}OH$[a,b]	3 M urea	25	10×10^{-5}	Schick (1965)
$C_9H_{19}C_6H_4(OC_2H_4)_{10}OH$[a,b]	6 M urea	25	24×10^{-5}	Schick (1965)
$C_9H_{19}C_6H_4(OC_2H_4)_{10}OH$[a,b]	3 M guanidinium Cl	25	14×10^{-5}	Schick (1965)
$C_9H_{19}C_6H_4(OC_2H_4)_{10}OH$[a,b]	1.5 M dioxane	25	10×10^{-5}	Schick (1965)
$C_9H_{19}C_6H_4(OC_2H_4)_{10}OH$[a,b]	3 M dioxane	25	18×10^{-5}	Schick (1965)
$C_9H_{19}C_6H_4(OC_2H_4)_{31}OH$[a,b]	H_2O	25	1.8×10^{-4}	Schick (1965)
$C_9H_{19}C_6H_4(OC_2H_4)_{31}OH$[a,b]	3 M urea	25	3.5×10^{-4}	Schick (1965)
$C_9H_{19}C_6H_4(OC_2H_4)_{31}OH$[a,b]	6 M urea	25	7.4×10^{-4}	Schick (1965)

Table 2.7a
(*Continued*)

Compound	Solvent	T (°C)	CMC (M)	Reference
$C_9H_{19}C_6H_4(OC_2H_4)_{31}OH$[a,b]	3 M guanidinium Cl	25	4.3×10	Schick (1965)
$C_9H_{19}C_6H_4(OC_2H_4)_{31}OH$[a,b]	3 M dioxane	25	5.7×10^{-4}	Schick (1965)
n-octyl β-D-glucoside	H_2O[a,b]	25	$2.5 \times^{-2}$	Shinoda (1961)
n-octyl β-D-glucoside	H_2O	25	2.2×10^{-3}	Shinoda (1961)
n-dodecyl β-D-glucoside	H_2O	25	1.9×10^{-4}	Shinoda (1961)
n-$C_6H_{13}[OCH_2CH(CH_3)]_2(OC_2H_4)_{9.9}OH$	H_2O	20	4.7×10^{-2}	Kucharski (1974)
n-$C_6H_{13}[OCH_2CH(CH_3)]_3(OC_2H_4)_{9.7}OH$	H_2O	20	3.2×10^{-2}	Kucharski (1974)
n-$C_6H_{13}[OCH_2CH(CH_3)]_4(OC_2H_4)_{9.9}OH$	H_2O	20	1.9×10^{-2}	Kucharski (1974)
n-$C_7H_{15}[OCH_2CH(CH_3)]_3(OC_2H_4)_{9.7}OH$	H_2O	20	1.1×10^{-2}	Kucharski (1974)
n-$C_{12}H_{25}N(CH_3)_3O$	H_2O	27	2.1×10^{-3}	Herrmann (1962)
Siloxane-based nonionics				
$(CH_3)_3SiO[Si(CH_3)_2O]_3$–$Si(CH_3)_2CH_2(C_2H_4O)_{8.2}CH_3$	H_2O	25	5.6×10^{-5}	Kanner (1967)
$(CH_3)_3SiO[Si(CH_3)_2O]_3$–$Si(CH_3)_2CH_2(C_2H_4O)_{12.8}CH_3$	H_2O	25	2.0×10^{-5}	Kanner (1967)
$(CH_3)_3SiO[Si(CH_3)_2O]_3$–$Si(CH_3)_2CH_2(C_2H_4O)_{17.3}CH_3$	H_2O	25	1.5×10^{-5}	Kanner (1967)
$(CH_3)_3SiO[Si(CH_3)_2O]_9$–$Si(CH_3)_2CH_2(C_2H_4O)_{17.3}CH_3$	H_2O	25	5.0×10^{-5}	Kanner (1967)

[a]Hydrophilic group not homogenous, but distribution of polyoxyethylene chains reduced by molecular distillation.
[b]Hydrophobic group equivalent to 10.5 C atoms in straight chain.

2.5.3.2. *Empirical*

Many empirical equations have been developed relating CMC of surfactants to structural units in them. For homologous straight-chain ionic surfactants (soaps, alkylsulfonates, alkylsulfates, alkylammonium chlorides and alkyltrimethyl-ammonium bromides) in aqueous medium, the following relationship between CMC and the number of carbon atoms (n) in the hydrophobic chain has been obtained (Somasundaran, 1964; Lin, 1971; Somasundaran and Fuerstenau, 1968; Bolden et al., 1983):

$$\log \mathrm{CMC} = A - Bn, \qquad (2.97)$$

where A is a constant for a particular ionic head at a given temperature and B is a constant for a particular hydrophobic chain, that is dependent on the temperature ($B = \log 2 \sim 0.3$ at 35°C). Eq. (2.97) shows a linear relationship between CMC and n for homologous compounds, as shown in Fig. 2.16 for cationic collectors and Fig. 2.17 for anionic collectors (Lin and Somasundaran, 1971; Fuerstenau et al., 1976). A and B values for the collectors and the frothers commonly used in nonsulfide flotation are shown in Table 2.8.

In the case of nonionic polyoxyethylenated surfactants, the CMC value is determined by the number of hydrocarbon groups, n, as well as oxyethylene groups, m, in the mole-

Table 2.7b
Aggregation number of some surfactant micelles

Compound	Solvent	T (°C)	Aggregation number	Reference
$C_8H_{17}SO_3^-Na^+$	H_2O	23	25	Tartar (1955)
$(C_8H_{17}SO_3^-)_2Mg^{2+}$	H_2O	23	51	Tartar (1955)
$C_{10}H_{21}SO_3^-Na^+$	H_2O	30	40	Tartar (1955)
$(C_{10}H_{21}SO_3^-)_2Mg^{2+}$	H_2O	60	103	Tartar (1955)
$C_{10}H_{21}SO_4^-Na^+$	H_2O	23	50	Tartar (1955)
$C_{12}H_{25}SO_3^-Na^+$	H_2O	40	54	Tartar (1955)
$(C_{12}H_{25}SO_3^-)_2Mg^{2+}$	H_2O	60	107	Tartar (1955)
$C_{12}H_{25}SO_4^-Na^{2+}$	H_2O	23	71	Tartar (1955)
$C_{14}H_{29}SO_3^-Na^+$	H_2O	60	80	Tartar (1955)
$C_{14}H_{29}SO_3^-Na^+$	0.01 M NaCl	23	138	Tartar (1955)
$C_{12}H_{25}N(CH_3)_3^+Br^-$	H_2O	23	50	Tartar (1955)
$[C_{12}H_{25}N(CH_3)_3^+]_2SO_4^{2-}$	H_2O	23	65	Tartar (1955)
$C_{12}H_{25}O(C_2H_4O)_6H$	H_2O	25	140	Balmbra (1962)
$C_{12}H_{25}O(C_2H_4O)_6H$	H_2O	25	400	Balmbra (1962)
$C_{12}H_{25}O(C_2H_4O)_6H$	H_2O	35	1400	Balmbra (1962)
$C_{12}H_{25}O(C_2H_4O)_6H$	H_2O	45	4000	Balmbra (1962)
$C_{12}H_{25}O(C_2H_4O)_8H^a$	H_2O	25	123	Becher (1961)
$C_{12}H_{25}O(C_2H_4O)_{12}H^a$	H_2O	25	81	Becher (1961)
$C_{12}H_{25}O(C_2H_4O)_{18}H^a$	H_2O	25	51	Becher (1961)
$C_{12}H_{25}O(C_2H_4O)_{23}H^a$	H_2O	25	40	Becher (1961)
$C_8H_{17}O(C_2H_4O)_6H$	H_2O	18	30	Balmbra (1964)
$C_8H_{17}O(C_2H_4O)_6H$	H_2O	30	41	Balmbra (1964)
$C_8H_{17}O(C_2H_4O)_6H$	H_2O	40	51	Balmbra (1964)
$C_8H_{17}O(C_2H_4O)_6H$	H_2O	60	210	Balmbra (1964)
$C_{10}H_{21}O(C_2H_4O)_6H$	H_2O	35	260	Balmbra (1964)
$C_{14}H_{29}O(C_2H_4O)_6H$	H_2O	35	7500	Balmbra (1964)
$C_{16}H_{33}O(C_2H_4O)_6H$	H_2O	34	16 600	Balmbra (1964)
$C_{16}H_{33}O(C_2H_4O)_6H$	H_2O	25	2430	Elworthy (1963)
$C_{16}H_{33}O(C_2H_4O)_7H$	H_2O	25	594	Elworthy (1963, 1964)
$C_{16}H_{33}O(C_2H_4O)_9H$	H_2O	25	219	Elworthy (1963)
$C_{16}H_{33}O(C_2H_4O)_{12}H$	H_2O	25	152	Elworthy (1963)
$C_{16}H_{33}O(C_2H_4O)_{21}H$	H_2O	25	70	Elworthy (1963)
$C_9H_{19}C_6H_4O(C_2H_4O)_{10}H^b$	H_2O	25	276	Schick (1962b)
$C_9H_{19}C_6H_4O(C_2H_4O)_{15}H^b$	H_2O	25	80	Schick (1962b)
$C_9H_{19}C_6H_4O(C_2H_4O)_{15}H^b$	0.5 M urea	25	82	Schick (1962b)
$C_9H_{19}C_6H_4O(C_2H_4O)_{15}H^b$	0.86 M urea	25	83	Schick (1962b)
$C_9H_{19}C_6H_4O(C_2H_4O)_{20}H^b$	H_2O	25	62	Schick (1962b)
$C_9H_{19}C_6H_4O(C_2H_4O)_{30}H^b$	H_2O	25	44	Schick (1962b)
$C_9H_{19}C_6H_4O(C_2H_4O)_{50}H^b$	H_2O	25	20	Schick (1962b)
$C_{10}H_{21}O(C_2H_4O)_8CH_3$	H_2O	30	83	Nakagawa (1960)
$C_{10}H_{21}O(C_2H_4O)_8CH_3$	H_2O + 2.3% *n*-decane	30	90	Nakagawa (1960)
$C_{10}H_{21}O(C_2H_4O)_8CH_3$	H_2O + 4.9% *n*-decane	30	105	Nakagawa (1960)
$C_{10}H_{21}O(C_2H_4O)_8CH_3$	H_2O + 3.4% *n*-decane	30	89	Nakagawa (1960)
$C_{10}H_{21}O(C_2H_4O)_8CH_3$	H_2O + 8.5% *n*-decane	30	109	Nakagawa (1960)
$C_{10}H_{21}O(C_2H_4O)_{11}CH_3$	H_2O	30	65	Nakagawa (1960)
$C_8H_{17}N^+(CH_3)_2CH_2COO^-$	H_2O	21	24	Tori (1960a)
$C_8H_{17}CH(COO^-)N^+(CH_3)_3$	H_2O	21	31	Tori (1960a)

Table 2.7b
(*Continued*)

Compound	Solvent	T (°C)	Aggregation number	Reference
a-Monocaprin	C_6H_6	–	42	Debye (1958)
a-Monolaurin	C_6H_6	–	73	Debye (1958)
a-Monomyristin	C_6H_6	–	86	Debye (1958)
a-Monopalmitin	C_6H_6	–	15	Debye (1958)
a-Monostearin	C_6H_6	–	11	Debye (1958)
n-$C_{12}H_{25}(OC_2H_4)_2OH$	C_6H_6	–	34	Becher (1960)
n-$C_{13}H_{27}(OC_2H_4)_6OH$	C_6H_6	–	99	Becher (1960)
n-$C_8H_{17}NH_3^+ \cdot C_2H_5COO^-$	C_6H_6	30	5 ± 1	Fendler (1973a)
n-$C_8H_{17}NH_3^+ \cdot C_2H_5COO^-$	CCl_4	30	3 ± 1	Fendler (1973a)
n-$C_8H_{17}NH_3^+ \cdot C_3H_7COO^-$	C_6H_6	30	3 ± 1	Fendler (1973a)
n-$C_8H_{17}NH_3^+ \cdot C_3H_7COO^-$	CCl_4	30	4 ± 1	Fendler (1973a)
n-$C_8H_{17}NH_3^+ \cdot C_5H_{11}COO^-$	C_6H_6	30	3 ± 1	Fendler (1973a)
n-$C_8H_{17}NH_3^+ \cdot C_5H_{11}COO^-$	CCl_4	30	5 ± 1	Fendler (1973a)
n-$C_8H_{17}NH_3^+ \cdot C_8H_{17}COO^-$	C_6H_6	30	3 ± 1	Fendler (1973a)
n-$C_8H_{17}NH_3^+ \cdot C_8H_{17}COO^-$	CCl_4	30	5 ± 1	Fendler (1973a)
n-$C_8H_{17}NH_3^+ \cdot C_{11}H_{23}COO^-$	C_6H_6	30	7 ± 1	Fendler (1973a)
n-$C_8H_{17}NH_3^+ \cdot C_{13}H_{27}COO^-$	C_6H_6	30	3 ± 1	Fendler (1973a)
n-$C_8H_{17}NH_3^+ \cdot C_{13}H_{27}COO^-$	CCl_4	30	3 ± 1	Fendler (1973a)
n-$C_4H_9NH_3^+ \cdot C_2H_5COO^-$	C_6H_6	–	4	Fendler (1973b)
n-$C_4H_9NH_3^+ \cdot C_2H_5COO^-$	CCl_4	–	3	Fendler (1973b)
n-$C_6H_{13}NH_3^+ \cdot C_2H_5COO^-$	C_6H_6	–	7	Fendler (1973b)
n-$C_6H_{13}NH_3^+ \cdot C_2H_5COO^-$	CCl_4	–	7	Fendler (1973b)
n-$C_8H_{17}NH_3^+ \cdot C_2H_5COO^-$	C_6H_6	–	5	Fendler (1973b)
n-$C_8H_{17}NH_3^+ \cdot C_2H_5COO^-$	CCl_4	–	5	Fendler (1973b)
n-$C_{10}H_{21}NH_3^+ \cdot C_2H_5COO^-$	C_6H_6	–	5	Fendler (1973b)
n-$C_{10}H_{21}NH_3^+ \cdot C_2H_5COO^-$	CCl_4	–	4	Fendler (1973b)

[a]Commercial material.

[b]Molecularly distilled commercial material.

cule $C_nH_{2n+1}(CH_2CH_2O)_mOH$. The following empirical relationship has been obtained for these compounds:

$$\log \mathrm{CMC} = A' + B'n, \tag{2.98}$$

where n is the number of hydrocarbon chains or oxyethylene units. For a given m, CMC has a linear relation with n and vice versa:

$$m = 6, \log \mathrm{CMC} = -0.5n + 1.03 \tag{2.99}$$

$$n = 10, \log \mathrm{CMC} = 0.08m - 3.37 \tag{2.100}$$

Constants, A' and B', for some nonionic surfactants used in flotation are shown in Table 2.9.

In flotation, CMC has been used for the following three purposes:

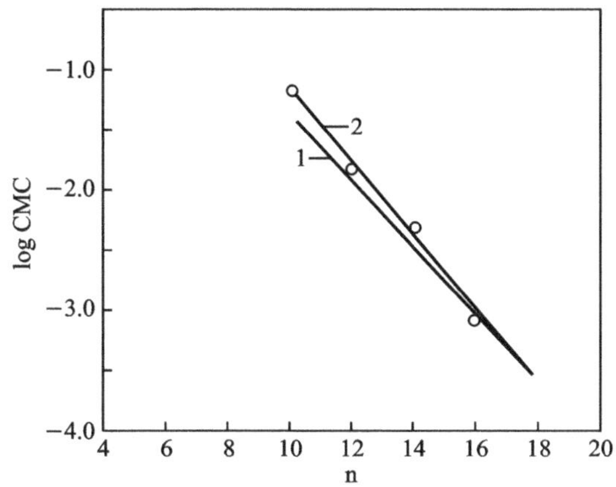

Fig. 2.16. log CMC of cationic collectors as a function of alkyl chain length n at 25°C: (1) RNH_2 and (2) $RN(CH_3)_3Br$ (Gaudin, 1934; Lin and Somasundaran, 1971).

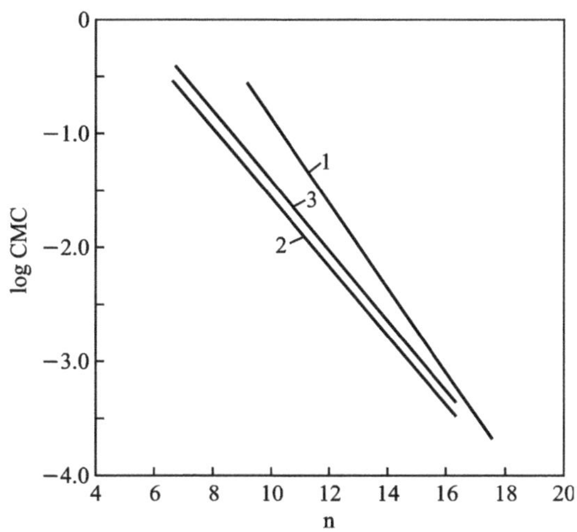

Fig. 2.17. log CMC of anionic collectors as a function of alkyl chain length n (Bolden, 1983; Lin, 1971; Fuerstenau, 1976) (at 25°C): (1) RCOONa (20°C), (2) $ROSO_3Na$ (25°C) and (3) RSO_3Na.

(1) Evaluation of type and performance of long-chain flotation reagents.
(2) Prediction of dosage range of flotation reagents.
(3) Factors that affect CMC and flotation.

Table 2.8
A and *B* values of the collectors and frothers commonly used in non-sulfide flotation

Flotagent	*A*	*B*	*T* (°C)	Ref.
RCOONa	2.41	0.341	20	Wang (1983)
RCOOK	1.92	0.290	25	Arai (1971)
RCOOK	2.03	0.292	45	Wang (1983)
RSO_3Na	1.64	0.300	25	Bustamante (1983)
RSO_3Na	1.59	0.294	40	Abramov (1961)
RSO_4Na	1.46	0.275	25	Jiang et al. (2001)
RSO_4Na	1.42	0.265	45	Abramov (1961)
RNH_2 HCl	1.25	0.295	25	Wang (1983)
$RN(CH_3)_3Br$	1.72	0.300	25	Lin (1971)
$RC_6H_4SO_3Na$	–	0.292	55	Lin and Somasundaran (1971)
$RN(CH_3)_3Cl$	1.64	0.286	25	Abramov (1961)

Table 2.9
Constants for the relationship: $\log C_{\mathrm{cmc}} = A' + B'n$

Surfactant series	*T* (°C)	A'	B'	Reference
n-$C_{12}H_{25}(OC_2H_4)_xOH$	23	−4.4	+0.016	Lange (1965)
n-$C_{12}H_{25}(OC_2H_4)_xOH$[a]	55	−4.8	+0.013	Schick (1962a)
p-1-$C_8H_{17}C_6H_4(OC_2H_4)_xOH$	25	−3.8	+0.029	Crook (1964)
$C_9H_{19}C_6H_4(OC_2H_4)_xOH$	25	−4.3	+0.020	Schick (1962b)
n-$C_{16}H_{33}(OC_2H_4)_xOH$	25	−5.9	+0.024	Elworthy (1962)

[a]Hydrophilic head not homogeneous, but distribution of polyoxyethylene chain lengths reduced by molecular distillation.

2.6. HYDROPHOBIC ASSOCIATION OF LONG-CHAIN SURFACTANTS

2.6.1. Hydrophobic association energy, ϕ

Hydrophobic association energy, ϕ, plays a major role in determining the extent of interactions among flotation reagents and mineral surfaces. In turn, the proportion of hydrophilic and hydrophobic groups in the flotation reagents is relevant in the determination of ϕ value.

The total association energy, ΔG, can be considered to be the sum of free energies for hydrophobic and hydrophilic groups, ΔG_{n} and ΔG_{p}.

$$\Delta G = \Delta G_{\mathrm{n}} + \Delta G_{\mathrm{p}}, \tag{2.101}$$

where

$$\Delta G_{\mathrm{n}} = n\phi \tag{2.102}$$

$$\Delta G_{\mathrm{p}} = \Delta G_{\mathrm{c}} + \Delta G_{\mathrm{e}} \tag{2.103}$$

Table 2.10
ϕ and K_0 values of some common surfactants determined from CMC data

Flotagent	T (°C)	K_0	ϕ/mol (–CH_2–)
RCOOK	25	0.56*	$-1.04RT$
RCOONa	20	0.56	$-1.23RT$
RSO_3Na	25	0–1	$-(0.64–1.27)RT$
RSO_3Na	45	0–1	$-(0.61–1.22)RT$
RSO_3Na	25	0–1	$-(0.69–1.38)RT$
RSO_3Na	40	0–1	$-(0.68–1.35)RT$
$RNH_2 \cdot HCl$	25	0–1	$-(0.63–1.36)RT$
$RN(CH_3)_3Br$	25	0.65*	$-1.13RT$
$RC_6H_4SO_3Na$	55	0–1	$-(0.67–1.34)RT$
$RN(CH_3)_3Cl$	25	0.65	$-1.09RT$

*Literature data.

ΔG_c is free energy change of chemical interactions and ΔG_e is free energy change due to electrostatic interaction.

For homologous compounds:

$$\phi = k_0 RT(M_{n+1} - M_n) \tag{2.104}$$

Various methods for determining value two are discussed below.

2.6.2. Methods for calculation of ϕ

2.6.2.1. *CMC method*

As can be seen from Eq. (2.95), log CMC exhibits a linear relationship with n. A log CMC vs. n diagram can be constructed if the CMC values of surfactants with different chain lengths are determined. Its slope will be:

$$B = -\phi/2.303(1 + K_0)RT \tag{2.105}$$

For given B and K, ϕ value can be calculated. Table 2.10 lists ϕ values for some common surfactants. It can be seen that ϕ values for fatty acids, alkylsulfates and alkyl ammonium chlorides calculated using CMC data range from $-0.6kT/CH_2$ to $1.3kT/CH_2$.

For nonionic surfactants also, log CMC and n are linearly related as per Eq. (2.96). Slope of the log CMC vs n plot in this case is $-\phi/RT$.

2.6.2.2. *Surface tension reduction method*

Since surface tension reduction depends on the replacement of solvent molecules at the interface by surfactant molecules, the efficiency of a surfactant in reducing the surface tension should reflect the concentration of the surfactant at the interface (C_s) relative to that of the bulk liquid phase (C_b). Generally, when the surface tension is reduced to 20 dynes/cm, the surfactant adsorption at the interface is close to its maximum. Adsorption at the interface can be explained by the following equation:

$$C_s/C_b = \exp(\Delta G/RT), \tag{2.106}$$

where ΔG is the free energy of transfer of a surfactant molecule from the interior of the liquid phase to the interface. At $\pi = 20$ dynes/cm,

$$\log(C_s/C_b) = \log(K_1/C_b)_{\pi=20} = -\phi'/2.303RT,$$

where ϕ is the free energy of transfer of a mole of surfactant molecules from the interior to the interface and K_1 is an association constant. $\log(1/C_b)_{\pi=20} = pC_{20}$ is thus directly related to the free energy change for the transfer process around maximum adsorption, ϕ, and is thus a more convenient and useful measure of the efficiency than $C_{b(\pi=20)}$ (to be used in adsorption later).

For a straight-chain surfactant of structure, $CH_3(CH_2)_nW$, where W is the hydrophilic portion of the molecule, ϕ can be divided into the free energy changes associated with the transfer of the terminal methyl group, $\phi(-CH_3)$, the $-CH_2-$ groups of the hydrocarbon chain, $\phi(-CH_2-)$ and the hydrophilic group, $\phi(-W)$, from the interior of the liquid phase to the interface at $\pi = 20$.

$$\log(1/C_b)_{\pi=20} = pC_{20} = [-\phi(-CH_2)/2.3RT]_m + [-\phi(-W)/2.3RT] + K \tag{2.107}$$

For a homologous series of straight-chain surfactants with the same hydrophilic group at the same temperature and under conditions where $\phi'(-W)$ is independent of the length of the hydrophobic group:

$$\log(1/C_b)_{\pi=20} = pC_{20} = [-\phi(-CH_2)/2.3RT]_m + K, \tag{2.108}$$

where K is a constant that indicates a larger free energy increase involved in the transfer of the hydrophilic head to the interface. Table 2.11 lists ϕ and K values for several homologous series of ionic surfactants in the absence of added inorganic electrolytes. The $\phi(-CH_2-)$ and K values of $-(0.65\text{–}0.70)RT$ in Table 2.11 mean that if the hydrophobic group of an ionic surfactant is increased by two $-CH_2-$ groups, the efficiency factor pC is increased by 0.56–0.6, and the same 20 dynes/cm reduction in surface tension can be obtained with a bulk concentration that is only 25–30% of the homologous surfactant with two less $-CH_2-$ groups.

2.6.2.3. ΔG *method*

The value of ϕ can also be obtained using the data for micellization of surfactants in which the surfactant is transferred from aqueous phase to an essentially hydrophobic phase. Table 2.12 lists transfer free energy data for some fatty acids and hydrocarbons.

2.6.2.4. *Zeta-potential method*

The effect of the chain length of alkyl ammonium acetates on the zeta-potential of quartz is shown in Fig. 2.18. With an increase in amine concentration, zeta-potential is reversed to positive values and with an increase in the chain length, n, the concentration required

Table 2.11
ϕ and K values of a few homologous series of ionic surfactants determined from surface tension reduction data (Zhao, 1984)

Flotagent	T (°C)	ϕ(–CH_2–)	K
RSO_4Na	25	$-0.70RT$	−1.12
RSO_3Na	25	$-0.70RT$	−1.26
$RC_6H_4SO_3Na$	70	$-0.65RT$	−1.27
RSO_4Na (at interface of heptane and water)	50	$-0.66RT$	−0.74
$RN(CH_3)_3Cl$	25	$-0.76RT$	−0.295
$R(OC_2H_4)_6OH$	25	$-0.99RT$	−0.08

Table 2.12
ϕ values of some fatty acid and hydrocarbons calculated from free energy transfer data (Nemethy and Scheraga, 1962; Mukerjee, 1965)

Fatty acid	C_8	C_{10}	C_{12}	C_{14}	C_{16}	ϕ kcal/mol (–CH_2–)
ΔG_m kcal/mol	−0.86	−2.54	−4.10	−5.80	−7.20	$-0.79 \approx -1.33RT$
Hydrocarbon	C_1	C_2	C_3	C_4		ϕ kcal/mol (–CH_2–)
ΔG_m kcal/mol	−2.83	−3.59	−4.90	−5.91		$-1.03 \approx -1.73RT$

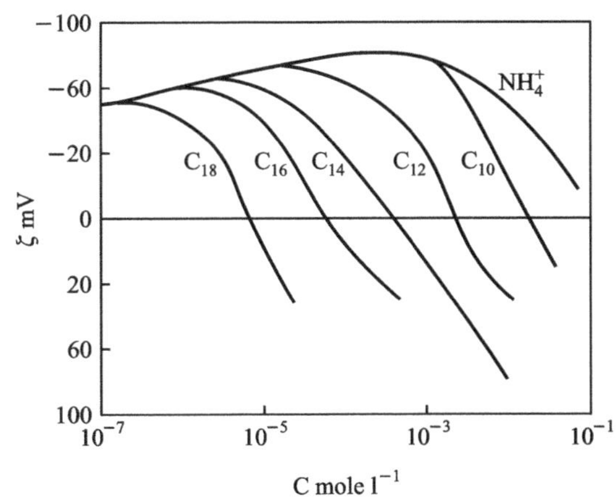

Fig. 2.18. Effect of chain length of alkyl ammonium acetates on zeta potential of quartz (Somasundaran, 1964).

for the reversal of zeta-potential is decreased consistently. A plot of log concentration for $\zeta = 0$ as a function of n is a linear line as shown in Fig. 2.19. On the basis of Stern–Graham equation, the slope of the line is equal to $-\phi/2.303RT$, and $\phi = -0.97RT$ or 2.3 kJ l/mol

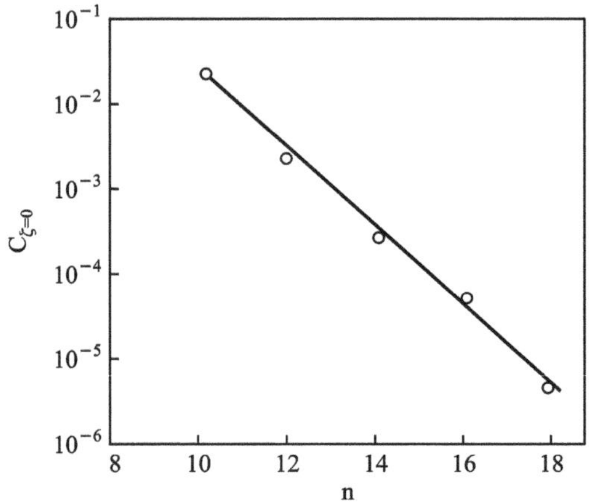

Fig. 2.19. Concentration of alkyl ammonium acetate required for $\zeta = 0$ as a function of alkyl chain length, n (Somasundaran, 1964).

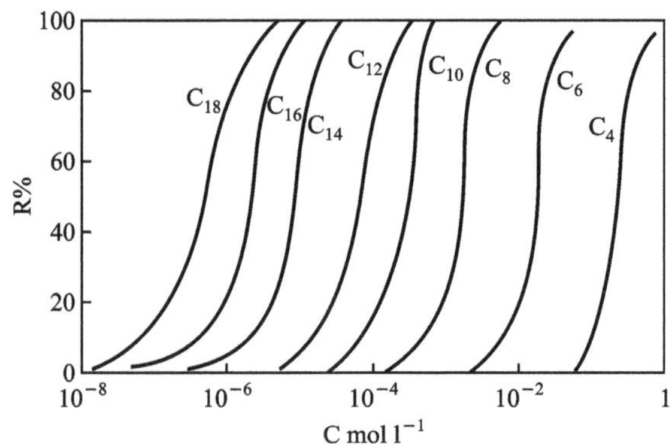

Fig. 2.20. Effect of chain length and concentration of amine acetates on the flotation recovery of quartz (Fuerstenau, 1964).

(CH_2), showing that the hydrophobic association energy of alkylamines when adsorbed on mineral surfaces is about 2.3 kJ/mol(–CH_2–).

2.6.2.5. *Flotation method*

The floatability of quartz using alkylamines is shown in Fig. 2.20. The concentration corresponding to the sharp increase in floatability is taken as C_{HM} (where C_{HM} is concentration at which hemimicelles form on the solid surface). $\log C_{HM}$–n is also linear with a slope equal to $-\phi/2.303RT$. ϕ was estimated to be $-1.0RT$ or 600 cal/mol(–CH_2–) in agreement with that obtained from the zeta-potential data.

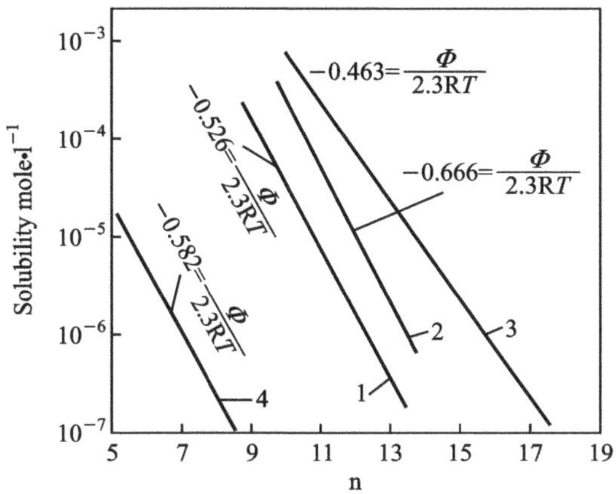

Fig. 2.21. Correlation between solubility of surfactants and alkyl chain length n: 1—fatty acids (25°C), 2—amines (22°C), 3—alkyl sulfonates (25°C), 4—alcohols (25°C) (Lin and Somasundaran, 1971).

2.6.2.6. *Solubility method*

Solubility of long-chain surfactants has been reported to have a linear relationship with the number of carbon atoms in hydrocarbon chain with a slope of $-\phi/2.303RT$. As shown in Fig. 2.21 (Lin and Somasundaran, 1971), ϕ values obtained for fatty acids, amines and alkyl sulfonates are $-1.45RT$, $-1.53RT$ and $-1.07RT$, respectively.

2.6.2.7. *Evaporation method*

In addition to the methods discussed above, ϕ values can also be estimated from data for heats of evaporation of homologous hydrocarbons. For example, when methane is evaporated from benzene solution, $\Delta G = -3.68$ kcal/mol whereas $\Delta G = 2.49$ kcal/mol for ethylene. ϕ calculated is equal to -1.19 kcal/mol or $-2.0RT$/mol (CH_2) group.

REFERENCES

Abramov, A.A., 1961. Izv. VUZ Tsvet. Metal. (1), 27–32.
Ananthapadmanabhan, K.P., et al., 1979. Trans. AIME 266, 2003.
Arai, H., et al., 1971. J. Coll. Inter. Sci. 37, 223.
Bolden, P.L., et al., 1983. J. Coll. Inter. Sci. 91 (2), 454.
Bustamante, H., et al., 1983. Trans. IMM Sec. C 92, C201–C207.
Chen, Z., Dai, M., 1985. Colloid Chemistry. High Education Press, Beijing.
Cook, M.A., Nixon, J.C., 1950. Phy. Chem. 54, 445–459.
Fuerstenau, D.W., et al., 1964. Trans. AIME 229, 321–325.
Fuerstenau, M.C., et al., 1968. Trans. AIME 241, 319.
Fuerstenau, M.C., et al., 1976. Flotation, Chapt. 5. In: Gaudin, A.M. (Ed.), Memorial Volume, vol. 1. AIME.
Fuerstenau, M.C., et al., 1977. Trans. AIME 26, 234.
Fuerstenau, M.C., et al., 1980. Metallic Ore Dressing Abroad (5), 57.
Gaudin, A.M., 1932. Flotation. McGraw-Hill Book Co., New York. 552 pp.
Gaudin, A.M., et al., 1934. Trans. Amer. Inst. Min. Metall. Engrs. 112, 319–347.
Hu, Y., (1984). Master Thesis, Central-South University.
Hu, Y., Wang, D., 1990. J. CSIMM 21 (2), 31–38.

Hu, Y., Wang, D., 1993. Nonferrous Metals 45 (2), 28–33.
Hu, Y., Liu, G., Wang, D., 1986. Nonferrous Metals 38 (2), 27–32.
Hu, Y.H., Chen, X.Q., Wang, Y.H., 2003. The Chinese Journal of Nonferrous Metals 1, 227–233.
Jiang, H., Hu, Y., Wang, D., 2001. Mining and Metallurgy Engineering 21 (2), 27–29.
Kulkarni, R.D., Somasundaran, P., 1980. Coll. Surf. 1, 387–405.
Li, Y., Li, C., 1983. Int. J. Miner. Process. 10, 205–218.
Lin, I.J., 1971. Trans. SME, AIME 250, 225–227.
Lin, I.J., Somasundaran, P., 1971. J. Coll. Inter. Sci. 37, 731.
Marabini, A.M., et al., 1984. In: Jones, M.J., Oblatt, R. (Eds.), Reagents in the Minerals Industry, pp. 125–136.
Mukerjee, P., 1958. J. Phy. Chem. 62, 1404.
Mukerjee, P., 1965. J. Phy. Chem. 69 (9), 2821.
Nemethy, G., Scheraga, H.A., 1962. J. Chem. Phys. 36 (2), 3401–3417.
Preston, W.C., 1948. J. Phy. Chem. 52, 84.
Pugh, R., Stenius, P., 1985. Int. J. Miner. Process. 15, 193–218.
Rosen, M., 2004. In: Surfactants and Interfacial Phenomena. Wiley-Interscience, Hoboken, pp. 113–120, 122–137; 1978 ibid.
Rosen, M., 1978. See Rosen, 2004.
Shah, D.O., et al., 1977. In: Shah, D.O., Schechter, R.S. (Eds.). Academic Press, New York, p. 106.
Somasundaran, P., 1964. J. Phy. Chem. 68, 12.
Somasundaran, P., 1976. Int. J. of Miner. Process. 3, 35.
Somasundaran, P., Ananthapadmanabhan, K.P., 1979a. In: Mittal, K.L. (Ed.), Solution Chemistry of Surfactants, vol. 2. Plenum Press New, York, pp. 777–799.
Somasundaran, P., Ananthapadmanabhan, K.P., 1979b. Trans. Indian Inst. Metals 32 (2), 177–191.
Somasundaran, P., Fuerstenau, D.W., 1968. Trans. AIME 241, 102.
Somasundaran, P., et al., 1984. J. Coll. Inter. Sci. 99 (1), 128–138.
Sugha, L., Kotrly, S., 1972. Solution Equilibria in Analytical Chemistry. Van Nostrand Reinhold Company, London.
Wang, D., 1983. Flotation Reagent—Fundamentals and Application. Metallurgy Industry Press, Beijing.
Wark, I.W., 1938. Principles of Flotation. Australasian Inst. Min. Metall., Melbourne. 346 pp.
Zhao, G., 1984. Phy. Chem. of the Surfactants. Beijing University Press, Beijing.

Chapter 3

Mineral–solution equilibria

Most minerals dissolve in aqueous solution to some extent, the extent of which is dependent, among other factors, upon the type and concentration of relevant chemical species in the solution. The dissolved mineral species can undergo further reactions such as hydrolysis, complexation, adsorption and even surface or bulk precipitation. The nature of the complex equilibria involving all such reactions will determine the interfacial properties of the particles and thus their flotation behavior. In the case of sparingly soluble minerals such as calcite, apatite, fluorite, etc., the effect of dissolved species on interfacial properties can be marked. These effects have been clearly established in the work on the calcite–apatite system (Hanna and Somasundaran, 1976; Amankonah, 1985; Somasundaran et al., 1985) where the surface conversion of apatite to calcite was shown to take place in their supernatants. The implications of such a conversion cannot be overemphasized. Surface conversion due to reactions between the dissolved species and the mineral surface can be predicted using the thermodynamic stability diagrams for heterogeneous mineral systems based on various equilibria for their dissociation (Hu and Wang, 1992; Hu et al., 1995, 1996a, 1996b).

3.1. DISSOLUTION OF MINERALS

3.1.1. Solubility product in water

Assuming that a mineral is in equilibrium with water and dissociation into cationic ions M^{n+} and anionic ions A^{m-} has taken place,

$$M_mA_n = m\ M^{n+} + n\ A^{m-} \tag{3.1}$$

Equilibrium constant of above reaction:

$$K_{sp} = [M^{n+}]^m[A^{m-}]^n \tag{3.2}$$

K_{sp} is called solubility product of the mineral and depends upon the ionic strength and on the temperature. If other reactions such as hydrolysis of cations M^{n+} and proton-addition of anions take place simultaneously, the solubility of minerals will be affected accordingly. A conditional solubility product can then be applied:

$$K'_{sp} = [M']^m[A']^n, \tag{3.2$'$}$$

where $[M']$ and $[A']$ are total concentrations of M and A in the saturated solution.

For the hydrolysis reaction of M^{n+},

$$[M'] = \alpha_M[M^{n+}], \tag{3.3a}$$

where α_M is ionization coefficient. For the proton-addition reaction of A^{m-},

$$[A'] = \alpha_A[A^{m-}] \tag{3.3b}$$

Then, substituting (3.3) into (3.2)

$$K^1_{sp} = [M^{n+}]^m[A^{m-}]^n\alpha^m_M\alpha^n_A = K_{sp}\alpha^m_M\alpha^n_A \tag{3.4}$$

For MA-type minerals,

$$K'_{sp} = K_{sp}\alpha_M\alpha_A \tag{3.5}$$

Assuming S_m is the solubility of mineral (mol/l)

$$[M'] = mS_m, \qquad [A'] = nS_m,$$

then

$$K'_{sp} = (mS_m)^m(nS_m)^n = K_{sp}\alpha^m_M\alpha^n_A \tag{3.6}$$

$$S_m = (K_{sp}\alpha^m_M\alpha^n_A/m^m n^n)^{1/(m+n)} \tag{3.7}$$

For MA-type minerals,

$$S_m = (K_{sp}\alpha_M\alpha_A)^{1/2} \tag{3.8}$$

Hydrolysis of cations and proton-addition of anions depend upon the type of the mineral. The solubility of various types of minerals can be calculated as follows.

3.1.1.1. *Sulfide minerals*

The hydrolysis of cations and proton-addition of anions should be considered in the calculation of solubility of sulfide minerals. In a saturated solution of sulfides, total concentration of S is:

$$[S'] = [S^{2-}] + [HS^-] + [H_2S]$$
$$= [S^{2-}](1 + K^H_1[H^+] + \beta^H_2[H^+]^2)$$

$$\alpha_s = 1 + \beta^H_1[H^+] + \beta^H_2[H^+]^2 \tag{3.9}$$

Total concentration of metallic ions is:

$$[M'] = [M^{n+}] + [M(OH)^{n-1}] + \cdots + [M(OH)^{N-K}_K]$$

$$\alpha_M = 1 + \beta_1[OH^-] + \beta_2[OH^-]^2 + \cdots + \beta_K[OH^-]^K \tag{3.10}$$

For M_mS_n,

$$K'_{sp} = K_{sp}\alpha^m_M\alpha^n_S \tag{3.11}$$

Take galena as an example:

$$PbS = Pb^{2+} + S^{2-} \qquad K_{sp} = 10^{-27.5} \tag{3.12}$$

Table 3.1
Solubility of some sulfide minerals in saturated solutions

Mineral	Chemical formula	Solubility (mol/l)	
		Calculated	Experimental
Covellite	CuS	3.6×10^{-15}	–
Chalcocite	Cu_2S	1.1×10^{-14}	–
Chalcopyrite	$CuFeS_2$	1.9×10^{-14}	–
Galena	PbS	7.9×10^{-11}	3.6×10^{-11}
Sphalerite	ZnS(α)	1.0×10^{-9}	1.47×10^{-9}
Sphalerite	ZnS(β)	1.6×10^{-9}	–
Pyrite	FeS_2	5.8×10^{-8}	–
Pyrrhotite	FeS	3.6×10^{-6}	–
Greenockite	CdS	1.23×10^{-10}	–
Linnaeite	CoS(α)	9.0×10^{-8}	–
Capillose	NiS(α)	8.1×10^{-7}	–
Argentite	Ag_2S	1.4×10^{-17}	–
Cinnabar	HgS	5.1×10^{-20}	–

Since the solubility of PbS is quite small, the change of pH due to proton-addition of S^{2-} can be ignored and pH of the solution can be considered to be about 7:

$$\alpha_S = 1 + 10^{13.9}[H^+] + 10^{20.92}[H^+]^2 = 1.63 \times 10^7$$

$$\alpha_{Pb} = 1 + 10^{6.3}[OH^-] + 10^{10.9}[OH^-]^2 + 10^{13.9}[OH^-]^3 = 1.2$$

$$K'_{sp} = 10^{-27.5} \times 1.63 \times 10^7 \times 1.2 = 6.19 \times 10^{-21}$$

$$[S_m] = (K'_{sp})^{1/2} = 7.87 \times 10^{-11}$$

Similarly, the solubilities of other sulfide minerals can be calculated using the data in Appendices C and D, and are shown in Table 3.1. It can be seen from the table that the solubilities of sulfide minerals in water are generally quite small. The calculated solubility is in good agreement with that reported in literature (Garrels and Christ, 1965; Linke, 1965).

3.1.1.2. *Oxide minerals*

Hydrolysis of metallic ions is the major reaction to be considered in the calculation of solubility of oxide minerals. For example, dissolution reaction of hematite in water is shown below:

$$Fe_2O_3 + 3H_2O = Fe^{3+} + 6OH^- \qquad K_{sp} = 10^{-42.7} \tag{3.13}$$

$$\alpha_{Fe} = 1 + 10^{11.81}[OH^-] + 10^{22.3}[OH^-]^2 + 10^{32.05}[OH^-]^3 + 10^{34.4}[OH^-]^4$$

In saturated solutions of Fe_2O_3, the concentration of OH^- can be assumed to be 1.0×10^{-7} because OH^- dissolved from Fe_2O_3 is negligible. Therefore, $\alpha_{Fe} = 1.12 \times 10^{11}$. The solubility of Fe_2O_3 is:

$$S_m = (K_{sp}\alpha_{Fe}/3^3)^{1/4} = 5.36 \times 10^{-9} \text{ mol/l}$$

Table 3.2
The solubility and pH values of some oxide minerals in saturated solutions

Mineral	Chemical formula	Solubility (mol/l)		pH_{sat}
		Calculated	Experimental	
Hematite	Fe_2O_3	5.4×10^{-9}	–	~7.00
Hydrargillite	$Al_2O_3{\cdot}3H_2O$	1.2×10^{-8}	–	~7.00
Zincite	ZnO	3.8×10^{-6}	–	8.85
Goethite	FeOOH	1.3×10^{-8}	–	~7.00
Cuprite	Cu_2O	2.2×10^{-7}	–	6.65
Tenorite	CuO	1.0×10^{-7}	–	7.01
Ostranite	ZrO_2	1.2×10^{-12}	–	~7.00
Rutile	TiO_2	7.9×10^{-12}	–	~7.00
Cassiterite	SnO_2	2.22×10^{-13}	–	~7.00
Lead oxide	PbO	2.45×10^{-5}	7.7×10^{-5}	9.69

Another example, hydrolysis reaction of ZnO, is shown below:

$$ZnO + H_2O = Zn^{2+} + 2OH^- \qquad K_{sp} = 10^{-16.66} \tag{3.14}$$

An approximate value for solubility of S, ZnO is

$$S = (K_{sp}/4)^{1/3} = 10^{-5.55} = 2.82 \times 10^{-6} \text{ mol/l}$$

Then,

$$[OH^-] = 2S = 5.64 \times 10^{-6} \text{ mol/l}, \ pH = 8.75$$

An approximate value of α_{Zn} can be calculated from $[OH^-]$ in saturated solution of ZnO.

$$\alpha_{Zn} = 1 + 10^{5.0}[OH^-] + 10^{11.1}[OH^-]^2 + 10^{13.6}[OH^-]^3 + 10^{14.8}[OH^-]^4 = 5.54$$

So that

$$K'_{sp} = 10^{-16.66} \times 5.54 = 10^{-15.92}$$

$$S_m = K'_{sp}/[OH^-]^2 = 3.78 \times 10^{-6}$$

Consider again $2S_m$ as a more accurate value for the concentration of $[OH^-]$. Calculation of α_{Zn} and K'_{sp}, then yields another S_m. More accurate values can be obtained for S_m and pH_{sat} by repeating such calculation many times. In this example, final S_m of ZnO is 3.5×10^{-6} mol/l and pH = 8.85.

Solubilities of some oxide minerals in saturated solutions and corresponding pH values can be calculated using the above method and the data are presented in Appendices C and D. The final values obtained are listed in Table 3.2 (Verwey and Overbeek, 1948). It can be seen from Table 3.2 that solubilities of oxides are larger than those of sulfides. The calculated values are in good agreement with those reported in literature (Garrels and Christ, 1965; Fuerstenau et al., 1976).

3.1.1.3. *Salt type minerals*

For salt type minerals, hydrolysis of cations and protonation of anions are taken into account in the calculation of their solubilities. Since the solubilities of these minerals are quite high, the protonation of dissolved anions from the mineral has a significant effect on pH.

3.1.1.3.1. *Carbonate minerals*

3.1.1.3.1.1. *Closed system* Take smithsonite, $ZnCO_3$, as an example:

$$ZnCO_3 = Zn^{2+} + CO_3^{2+}$$

$$K_{sp} = 10^{-10.0} \tag{3.15}$$

Hydrolysis of CO_3^{2-}: $CO_3^{2-} + H_2O = HCO_3^- + OH^-$

$$[OH^-][HCO_3^-]/[CO_3^{2-}][H_2O] = K_w/K_{a2} \tag{3.16}$$

Approximately, $[OH^-] = [HCO_3^-]$, $[CO_3^{2-}] = 10^{-5}$ (calculated from K_{sp}).

Thus, in equilibrium,

$$[CO_3^{2-}] = 10^{-5} - [OH^-] \tag{3.17}$$

Substituting (3.17) into (3.16),

$$[OH^-]^2 + 10^{-3.67}[OH^-] - 10^{-8.67} = 0$$

$$[OH^-] = 1.9 \times 10^{-5} \text{ mol/l}, \text{ pH } = 9.28$$

Then

$$\alpha_{Zn} = 1 + 10^{5.0} \times 1.9 \times 10^{-5} + 10^{11.1} \times (1.9 \times 10^{-5})^2 + 10^{13.6} \times (1.9 \times 10^{-5})^3 + 10^{14.8} \times (1.9 \times 10^{-5})^4 = 48.35$$

$$\alpha_{CO_3^{2-}} = 1 + 10^{10.33} \times 10^{9.28} + 10^{16.68} \times (10^{-9.28})^2 = 12.22$$

$$K'_{sp} = 10^{-10.0} \times 48.35 \times 12.22 = 5.9 \times 10^{-8}$$

$$S_m = (K'_{sp})^{1/2} = 2.43 \times 10^{-4} \text{ mol/l}$$

The concentration of free CO_3^{2-} is:

$$[CO_3^{2-}] = S_m/12.22 = 1.99 \times 10^{-5} \text{ mol/l}$$

Using (3.16), a more accurate value for $[OH^-]$ can be obtained as follows:

$$\frac{[OH^-]^2}{1.99 \times 10^{-5} - [OH^-]} = 10^{-14.0} \times 10^{10.33} = 10^{-3.67}$$

$$[OH^-] = 1.83 \times 10^{-5}, \qquad \text{pH } = 9.26$$

So $\alpha_{Zn} = 45.0$, $\alpha_{CO_3^{2-}} = 12.75$

$$S_m = (10^{-10} \times 4.31 \times 542.7)^{1/2} = 2.4 \times 10^{-4} \text{ mol/l}$$

The final value of S_m obtained using the step-by-step method is:

$$S_m = 2.4 \times 10^{-4}, \qquad pH_{sat} = 9.26$$

3.1.1.3.1.2. *Open systems* In open systems, due to the effect of CO_2 in the atmosphere on the dissolution of carbonates, the equilibrium is treated using equations (3.17)–(3.19):

$$[CO_3^{2-}][H^+]^2 = 10^{-21.64} \tag{3.18'}$$

Initial concentration of CO_3^{2-} is 10^{-5}, obtained from $K_{sp,ZnCO_3}$.
From (3.18′), pH = 8.32.

$$\alpha_{Zn} = 1 + 10^{5.0} \times 10^{-5.68} + 10^{11.1} \times (10^{-5.68})^2 + 10^{13.6} \times (10^{-5.68})^3 + 10^{14.8} \times (10^{-5.68})^4 = 1.76$$

$$\alpha_{CO_3^{2-}} = 1 + 10^{10.33} \times 10^{-8.32} + 10^{16.68} \times (10^{-8.32})^2 = 104.4$$

$$K'_{sp} = 10^{-10} \times 1.76 \times 104.4 = 1.84 \times 10^{-8}$$

$$S_m = (K'_{sp})^{1/2} = 1.3555 \times 10^{-4} \text{ mol/l}$$

A more accurate value for the concentration of free CO_3^{2-} ions is:

$$[CO_3^{2-}] = S_m/104.4 = 1.298 \times 10^{-6}$$

The final values for S_m and pH set obtained using the iterative method are:

$$S_m = 2.24 \times 10^{-4}, \qquad pH_{sat} = 7.65$$

pH_{sat} obtained in open systems is closer to the actual one observed in real systems.

3.1.1.3.2. *Phosphate minerals* Dissolution equilibria for aluminum phosphate can be written as follows:

$$AlPO_{4(s)} = Al^{3+} + PO_4^{3-} \qquad K_{sp} = 10^{-18.24} \tag{3.19}$$

$$PO_4^{3-} + H_2O = OH^- + HPO_4^{2-}$$

$$[OH^-][HPO_4^{2-}]/[PO_4^{3-}][H_2O] = K_w/K_{a2} = 10^{-14}/10^{-12.35} = 10^{-1.65} \tag{3.20}$$

Approximate initial value for PO_4^{3-} concentration obtained from K_{sp} is $10^{-9.12}$:

$$[PO_4^{3-}] = 10^{-9.12} - [OH^-]$$

Thus

$$[OH^-]^2 + 10^{-1.65}[OH^-] - 10^{-10.77} = 0 \tag{3.21}$$

$$[OH^-] = 10^{-8.8}, \qquad pH_{sat} = 5.2$$

$$\alpha_{PO_4} = 1 + 10^{12.35}[H^+] + 10^{19.55}[H^+]^2 + 10^{21.7}[H^+]^3 = 1.43 \times 10^9$$

$$\alpha_{Al} = 1 + 10^{9.9}[OH^-] + 10^{18.7}[OH^-]^2 + 10^{27.0}[OH^-]^3 + 10^{33.0}[OH^-]^4 = 19.2$$

$$K'_{sp} = 10^{-18.2} \times 19.2 \times 1.43 \times 10^9 = 1.5 \times 10^{-8}$$

$$S_m = 1.26 \times 10^{-4} \text{ mol/l}$$

3.1.1.3.3. *Silicate minerals* The dissolution of complex silicate minerals in aqueous solutions takes place in steps. For example,

Dissolution of kaolinite is in two steps:

(i) $H_4Al_2Si_2O_{9(s)} + 5H_2O = Al_2O_3 \cdot 3H_2O + 2H_4SiO_{4(aq)}$ (3.22)

(ii) In acidic solutions,

$$Al_2O_3 \cdot 3H_2O + 6H^+ = 2Al^{3+} + 6H_2O \quad (3.23a)$$

In basic solutions,

$$Al_2O_3 \cdot 3H_2O + 2OH^- = 2Al(OH)_4^- \quad (3.23b)$$

Total reactions: in acidic solutions,

$$1/2H_4Al_2Si_2O_{9(s)} + 3H^+ = Al^{3+} + H_4SiO_{4(aq)} + 1/2H_2O \qquad K_1 = 10^{3.35} \quad (3.24a)$$

In basic solutions,

$$1/2H_4Al_2Si_2O_{9(s)} + 5/2H_2O + OH^- = Al(OH)_4^- + H_4SiO_{4(aq)}$$
$$K_2 = 10^{-5.7} \quad (3.24b)$$

The correlation between solubility of kaolinite and pH is:

$$\log[Al]_T = \log K_1 + \log \alpha_{Al} - \log[H_4SiO_4] - 3pH \quad (3.25)$$

In pure water, assuming pH = 7

$$[Al]_T = 1.29 \times 10^{-7} \text{ mol/l}$$

3.1.1.3.4. *Sulfate minerals* Since sulfuric acid is a strong acid, only hydrolysis of cations is considered for the calculation of solubility of the sulfate minerals. For example:

$$PbSO_4 = Pb^{2+} + SO_4^{2-} \qquad K_{sp} = 10^{-6.20} \quad (3.26)$$

$$Pb^{2+} + H_2O = PbOH^+ + H^+$$

$$[H^+][PbOH^+]/[Pb^{2+}] = K_w \cdot K_1 \quad (3.27)$$

Initially, $[Pb^{2+}] = 10^{-3.1}$, and assuming $[H^+] = [PbOH^+]$

$$[H^+]^2 + 10^{-7.7}[H^+] - 10^{-10.8} = 0 \quad (3.28)$$

Then $[H^+] = 10^{-5.4}$, pH = 5.4

$$\alpha_{Pb} = 1 + 10^{6.3}[OH^-] + 10^{10.9}[OH^-]^2 + 10^{13.9}[OH^-]^3 = 1$$
$$S_m = (K_{sp})^{1/2} = 10^{-3.1} = 7.9 \times 10^{-4} \text{ mol/l}$$

Solubility of various salt type minerals in water and the pH of their solutions can be calculated from the data in Appendices C and D using the approach discussed previously. The

Table 3.3
Point of zero charge (PZC) of some oxide and silicate minerals

Augite		2.7
Bentonite		<3.0
Beryl		3.1, 3.3, 4.4
Biotite		0.4
Cassiterite		4.5
Chromite		5.6, 7.0, 7.2
Chrysocolla		2.0
Corundum		9.0, 9.4
Cummingtonite		5.2
Cuprite		9.5
Diopside		2.8
Garnet		4.4
Goethite		6.7
Hematite		5.0, 6.0, 6.7
Kaolinite		3.4
Magnetite		6.5
Pyrolusite		5.6, 7.4
Quartz		1.8
Rhodonite		2.8
Rutile		6.7
Tourmaline		4.0
Zircon		5.8
Barite, $BaSO_4$	pBa	6.7
Calcite, $CaCO_3$	pH	9.5*
Fluorapatite, $Ca_5(PO_4)_3(F,OH)$	pH	6*
Fluorite, CaF_2	pCa	3
Hydroxyapatite, $Ca_5(PO_4)_3(OH)$	pH	7*
Scheelite, $CaWO_4$	pCa	4.8
Silver chloride, AgCl	pAg	4
Silver iodide, AgI	pAg	5.6
Silver sulfide, Ag_2S	pAg	10.2

*From the hydrolysis equilibria and solubility data, the activities of the other potential-determining ions can be calculated.

results obtained are shown in Table 3.3. It can be seen that the solubilities of various salt type minerals are generally larger than those of the corresponding sulfide and oxide minerals. The pH value of the saturated solutions are either acidic or basic, suggesting that the dissolution and flotation of salt type minerals are much more complicated than those of sulfide and oxide minerals.

3.2. HYDROLYSIS EQUILIBRIA OF IONIC SPECIES

Presence of even trace amounts of sparingly soluble minerals can lead to fairly high concentrations of active inorganic species such as polyvalent metallic ions. These species have a significant effect on interfacial properties of minerals in solutions. Therefore, hydrolysis equilibria of inorganic species become the basis in flotation mechanisms.

3.2.1. Hydrolysis equilibria and log C–pH diagram

Hydrolysis reactions of polyvalent metallic ions take place in solutions to form various types of hydroxy complexes. the concentration of each complex can be calculated based on the solution equilibria of the cations.

3.2.1.1. *Homogeneous systems*

The hydrolysis equilibria of metallic cations are shown as follows.

$$\begin{aligned}
&M^{m+} + OH^- = MOH^{m-l} && \beta_1 = [MOH^{m-1}]/[M^{m+}][OH^-] \\
&M^{m+} + 2OH^- = M(OH)_2^{m-2} && \beta_2 = [M(OH)_2^{m-2}]/[M^{m+}][OH^-]^2 \\
&\dots && \dots \\
&M^{m+} + nOH^- = M(OH)_n^{m-n}, && \beta_n = [M(OH)_n^{m-n}]/[M^{m+}][OH^-]^n
\end{aligned} \tag{3.29}$$

where $\beta_1, \beta_2, \dots, \beta_n$ are cumulative stability constants. Assume $[M']$ represents the sum of concentrations of each component in the solution.

Thus,

$$\begin{aligned}
[M'] &= [M^{m+}] + [MOH^{m-1}] + [M(OH)_2^{m-2}] + \cdots + [M(OH)_n^{m-n}] \\
&= [M^{m+}](1 + \beta_1[OH^-] + \beta_2[OH^-]^2 + \cdots + \beta_n[OH^-]^n)
\end{aligned} \tag{3.30}$$

Define α_M as the reaction coefficient

$$\alpha_M = [M']/[M^{m+}] = 1 + \beta_1[OH^-] + \beta_2[OH^-]^2 + \cdots + \beta_n[OH^-]^n \tag{3.31}$$

The concentration of each component is

$$[M^{m+}] = [M']/\alpha_M = [M']/(1 + \beta_1[OH^-] + \beta_2[OH^-]^2 + \cdots + \beta_n[OH^-]^n) \tag{3.32}$$

$$\begin{aligned}
&\log[M^{m+}] = \log[M'] - \log(1 + \beta_1[OH^-] + \beta_2[OH^-]^2 + \cdots + \beta_n[OH^-]^n) \\
&\log[MOH^{m-1}] = \log\beta_1 + \log[M^{m+}] + \log[OH^-] \\
&\log[MOH^{m-2}] = \log\beta_2 + \log[M^{m+}] + 2\log[OH^-] \\
&\dots \\
&\log[M(OH)_n^{m-n}] = \log\beta_n + \log[M^{m+}] + n\log[OH^-]
\end{aligned} \tag{3.33}$$

3.2.1.2. *Heterogeneous system*

When metallic ions form precipitates in solutions, $M(OH)_{m(s)}$ is in equilibrium with other species:

$$\begin{aligned}
&M(OH)_{m(s)} = M^{m+} + mOH^- && K_{s0} = [M^{m+}][OH^-]^m \\
&M(OH)_{m(s)} = M(OH)^{m-1} + (m-1)OH^- && K_{sl} = [M(OH)^{m-1}][OH^-]^{m-1}
\end{aligned}$$

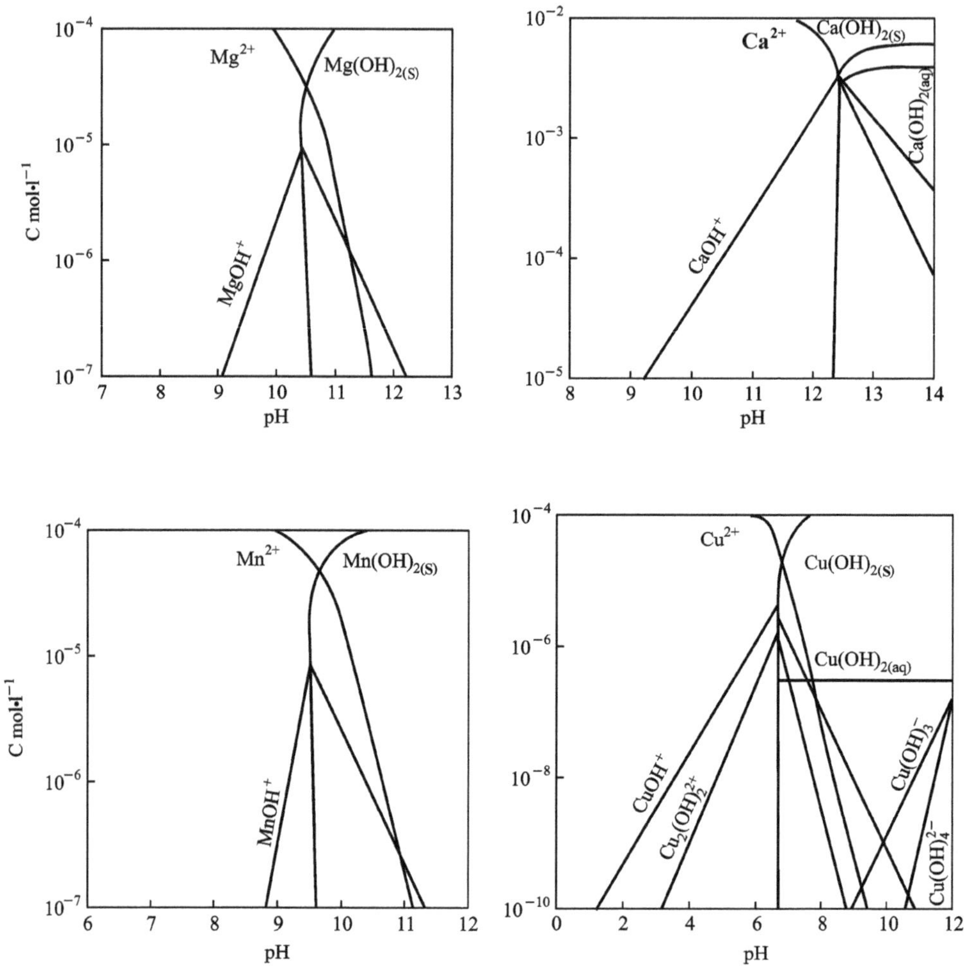

Fig. 3.1a. The log C–pH diagram for hydrolysis species of Mg, Ca, Mn and Cu ions (Fuerstenau, 1976; Morgan, 1981).

...

$$M(OH)_{m(s)} = M(OH)_n^{m-n} + (m-n)OH^- \qquad K = [M(OH)_n^{m-n}][OH^-]^{m-n} \tag{3.34}$$

The concentration of each species is given by

$$\log[M^{m+}] = \log K_{s0} - m\log[OH^-]$$

$$\log[M(OH)^{m-1}] = \log K_{s1} + (1-m)\log[OH^-]$$

...

$$\log[M(OH)_n^{m-n}] = \log K_{sn} + (n-m)\log[OH^-] \tag{3.35}$$

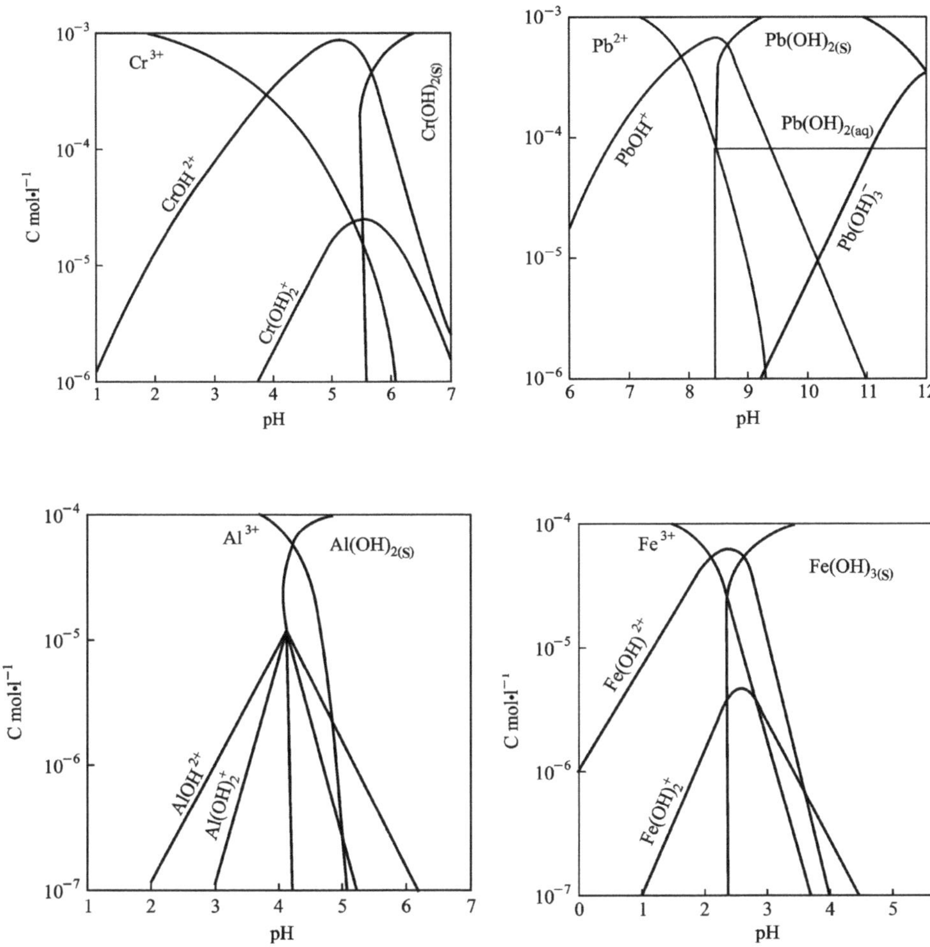

Fig. 3.1b. The log C–pH diagram for hydrolysis species of metallic ions (Fuerstenau, 1976; Morgan, 1981).

The log C–pH diagrams of species of metallic ions such as Ca, Mg, Fe, Cu, Pb, Cr, Al and Mn are shown in Figs. 3.1a and 3.1b. In each system, metallic ions, hydroxyl ions and hydroxides are present in different pH regions. These species play different roles in determining the flotation of the minerals.

3.2.2. Activation of flotation by hydroxy ions

Polyvalent metallic ions can adsorb specifically on oxides and silicates; this phenomenon occurs when the cation involved hydrolyzes to its first hydroxy complex or its dioxide precipitates on the particle surface. Unintentional activation of silica by dissolved metallic species in anionic flotation circuits is a typical example (Hanna and Somasundaran, 1976).

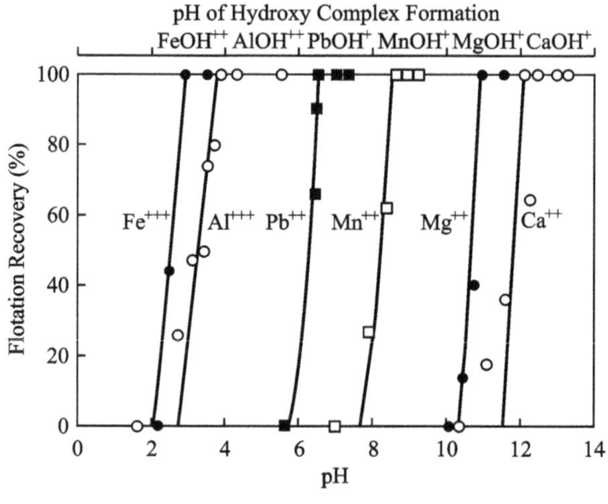

Fig. 3.2. Minimum flotation edges of quartz as a function of pH (1×10^{-4} M sulfonate and 1×10^{-4} M metallics ions) (Fuerstenau, 1976).

Multivalent cations will adsorb minerals such as silica and this process exhibits a sharp dependence on pH.

In general, adsorption of cations increases markedly at pH values just below that required for the precipitation of the corresponding metal hydroxide. In the case of quartz, since its solubility is very limited and the only cation comprising the mineral is silicon, flotation is obtained only after metal ions are added to the system in a pH range in which hydrolysis to first hydroxy complexes in bulk but possibly precipitation of the hydroxide on the surface occurs. The flotation edges outlining minimum values of pH at which flotation is obtained in the presence of 1×10^4 of various metal ions are shown in Fig. 3.2. The range of pH in which the flotation is obtained in the distribution diagrams (Figs. 3.1a and 3.1b) with these metal ions can estimated. The absence of flotation above the upper flotation edge and below the lower flotation edge is attributed to the insufficient amount of hydroxy complexes for flotation.

This activation has been attributed to the chemisorption of the first-hydroxy species which exist in significant amounts under those pH conditions (Fuerstenau, 1976). A typical example, viz., the adsorption of Ca species on silica, is shown in Fig. 3.3 (Clark and Cooke, 1968). The sharp increase in Ca adsorption around pH 10 has been attributed to the chemisorption of $CaOH^+$ which begins to form in significant amounts in the bulk around this pH. However, it is to be noted that since the adsorption can also be attributed to $Ca(OH)_2$ precipitation on the surface. Interfacial concentration of Ca^+ is higher that in the bulk solution Ca^{2+} (3×10^{-10} mol/cm^2) in the pH region of 4 to 10. Assuming that the thickness of the adsorbed layer is 10 or 4 Å (contact adsorption), the pH value for precipitation of $Ca(OH)_2$ is 11.5 or 10.96 respectively. Note that the surface region needs then to get up to only about pH 11 for $Ca(OH)_2$ precipitation to occur in that region. In the absence of any added Ca, the surface pH of silica can be expected to be lower than its bulk value.

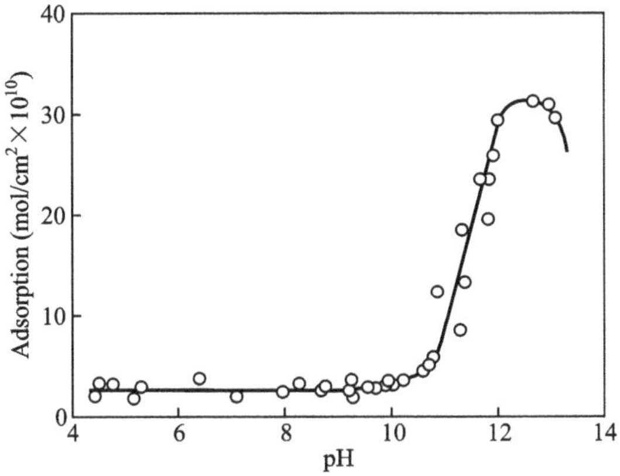

Fig. 3.3. Adsorption of calcium species on quartz as a function of pH from solution (100 ppm Ca^{2+}) (Clark and Cooke, 1968).

However, our estimates based on available information on the charge density of silica (40 C/cm^2 at pH 11 and ionic strength 10^{-2} kmol/m^3) suggest that in the present case, the surface pH will not be significantly different from the bulk pH, since 3×10^{-10} mol/cm^2 of calcium is reported to adsorb at this pH. Thus, it can be seen that the sharp increase in adsorption coincides with the conditions for the surface precipitation of $Ca(OH)_2$. Adsorption of a number of other metallic cations has been analyzed along the lines discussed above and in all such cases the sharp increase in adsorption at a particular pH can be accounted for by the surface precipitation of the hydroxide of the adsorptive.

Similar phenomena have been observed in numerous other systems such as those of aluminum quartz (Ananthapadmanabhan and Somasundaran, 1985), nickel-quartz (Nagaraj and Somasundaran, 1981), cupric-chrysocolla (Latiner, 1952), manganous-rhodonite (Butler, 1967) and ferrous-chromite.

3.3. SOLUTION EQUILIBRIA OF DISSOLVED MINERAL SPECIES

3.3.1. log C–pH diagram

When mineral particles are contacted with water, they will undergo dissolution, the extent of which is dependent upon the type and concentration of chemicals in solution. The dissolved mineral species can undergo further reactions such as hydrolysis, complexation, adsorption and even surface or bulk precipitation. The complex equilibria involving all such reactions can be expected to determine the interfacial properties of the particles and their flotation behavior. The concentration of each dissolved mineral species can be calculated from various solution equilibria of the minerals. The calculated results are plotted as log C–pH diagram. The equilibria in selected salt-type mineral systems with special reference to calcite and apatite are examined below.

Two relevant situations are considered: (i) closed system: system closed to atmospheric CO_2; (ii) open system: system open to atmospheric CO_2.

When calcite is brought in contact with water, its dissolution will be followed by pH dependent hydrolysis and complexation of the dissolved species. The following are the equilibria that will control the system (Amankonah, 1985; Somasundaran and Wang, 1984).

	K
$CaCO_{3(s)} = Ca^{2+} + CO_3^{2-}$	$10^{-8.4}$
$CO_3^{2-} + H^+ = HCO_3^-$	$10^{10.3}$
$HCO_3^- + H^+ = H_2CO_3$	$10^{6.3}$
$CO_{2(g)} + H_2O = H_2CO_3$	$10^{-1.5}$
$Ca^{2+} + HCO_3^- = CaHCO_3^+$	$10^{0.8}$
$Ca^{2+} + CO_3^{2-} = CaCO_{3(aq)}$	$10^{3.3}$
$Ca^{2+} + H_2O = CaOH^+ + H^+$	$10^{-12.9}$
$Ca^{2+} + 2H_2O = Ca(OH)_2 + 2H^+$	$10^{-22.8}$

While the activities can fairly easily be calculated in the case of the open systems due to the fixed pressure of atmospheric CO_2 at $10^{-3.5}$ atm, calculations for the closed systems are relatively tedious. A computer program is developed for calculating the activities of various dissolved species from available thermodynamic data, taking into consideration the effects of other dissolved species and solid phases in equilibrium with the system. Figs. 3.4a and 3.4b show the distribution of various species as a function of pH for the closed and open systems, respectively. The marked effect of atmospheric CO_2 on the solubility of calcite can be seen by comparing the data in these diagrams. This effect is particularly noticeable for Ca^{2+}, HCO^{3-} and CO_3^{2-} activities. Since Ca plays an important role in calcite flotation systems, the possible role of CO_2 as well as other interfacial species on flotation due to this effect should be noted.

The equilibria that control the dissolution of apatite in water are given below:

$$
\begin{aligned}
&Ca^{2+} + OH^- = CaOH^+ && HPO_4^{2-} + H_2O = H_2PO_4^- + OH^- \\
&CaOH^+ + OH^- = Ca(OH)_2 && H_2PO_4^- + H_2O = H_3PO_4 + OH^- \\
&Ca(OH)_{2(aq)} = Ca(OH)_{2(s)} && Ca^{2+} + HPO_4^{2-} = CaHPO_4 \\
&PO_4^{3-} + H_2O = HPO_4^{2-} + OH^- && Ca^{2+} + H_2PO_4^- = CaH_2PO_4^+ \qquad (3.36)
\end{aligned}
$$

As in the case of calcite, the activities of various dissolved species have been calculated at different pH values for other minerals using the relevant chemical equilibria and stoichiometric restrictions. The distribution of the activities of dissolved species of hydroxy apatite is given in Figs. 3.5a and 3.5b for open and closed systems, respectively. If the system is brought to equilibrium with atmospheric carbon dioxide, the distribution will follow that shown in Fig. 3.5a. For the closed system, Ca species are the most predominant below about pH 12.5. The predominant phosphate species are HPO_4^{2-}, $H_2PO_4^-$ and PO_4^{3-}

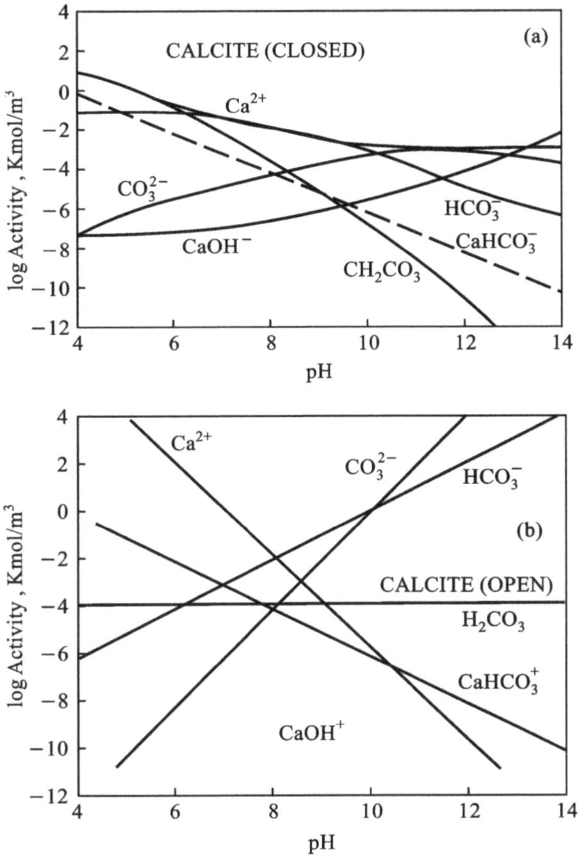

Fig. 3.4. Species distribution diagrams for calcite–water systems: (a) closed and (b) open.

depending on the pH. By comparing the data in Fig. 3.5, it can be seen that, below about pH 9, the effect of atmospheric CO_2 on Ca activity is not as drastic in this case as for calcite (Fig. 3.4). Above this pH, marked changes are observed: while the Ca activity decreases sharply with increase in pH, an increase is observed in the activities of HPO_4^{2-} and PO_4^{2-} species (Fig. 3.5a). The sharp decrease in the Ca activity with increase in the phosphate solubility under high pH conditions is due to the precipitation of $CaCO_3$.

Tungsten minerals such as wolframite and scheelite are other examples of sparingly soluble mineral systems. The major solution equilibria of wolframite (Mn, Fe)WO_4 and scheelite $CaWO_4$ are as follows (Hu and Wang, 1985):

(a)

$$MnWO_{4(s)} = Mn^{2+} + WO_4^{2-} \qquad pK_{spl} = 8.85$$

$$Mn(OH)_{2(s)} = Mn^{2+} + 2OH^- \qquad pK_{sl} = 12.6$$

$$WO_{3(s)} + H_2O = 2H^+ + WO_4^{2-} \qquad pK_{s0} = 10.95$$

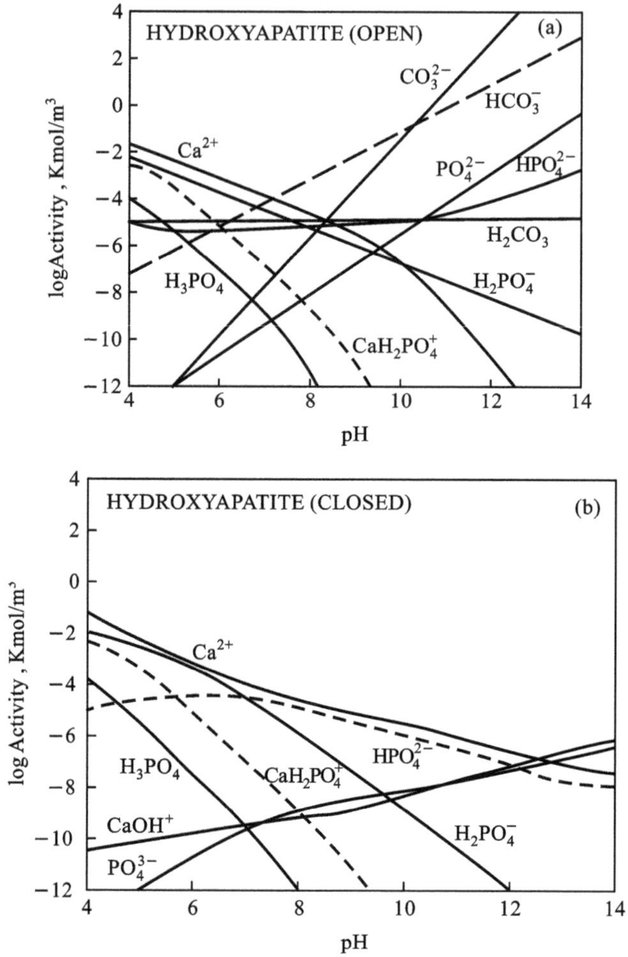

Fig. 3.5. Species distribution diagrams for hydroxyapatite–water systems: (a) open and (b) closed.

(b)

$$FeWO_{4(s)} = Fe^{2+} + WO_4^{2-} \qquad pK_{sp2} = 11.04$$

$$Fe(OH)_{2(s)} = Fe^{2+} + 2OH^- \qquad pK_{s2} = 14.95$$

(c)

$$CaWO_{4(s)} = Ca^{2+} + WO_4^{2-} \qquad pK_{sp3} = 9.3$$

$$Ca(OH)_{2(s)} = Ca^{2+} + 2OH^- \qquad pK_{s3} = 5.22 \qquad (3.37)$$

The species distribution diagrams calculated from these equilibria are shown in Figs. 3.6a–3.6c. For wolframite, both Mn^{2+} and Fe^{2+} are dominant species for pH < 2.8. The mineral

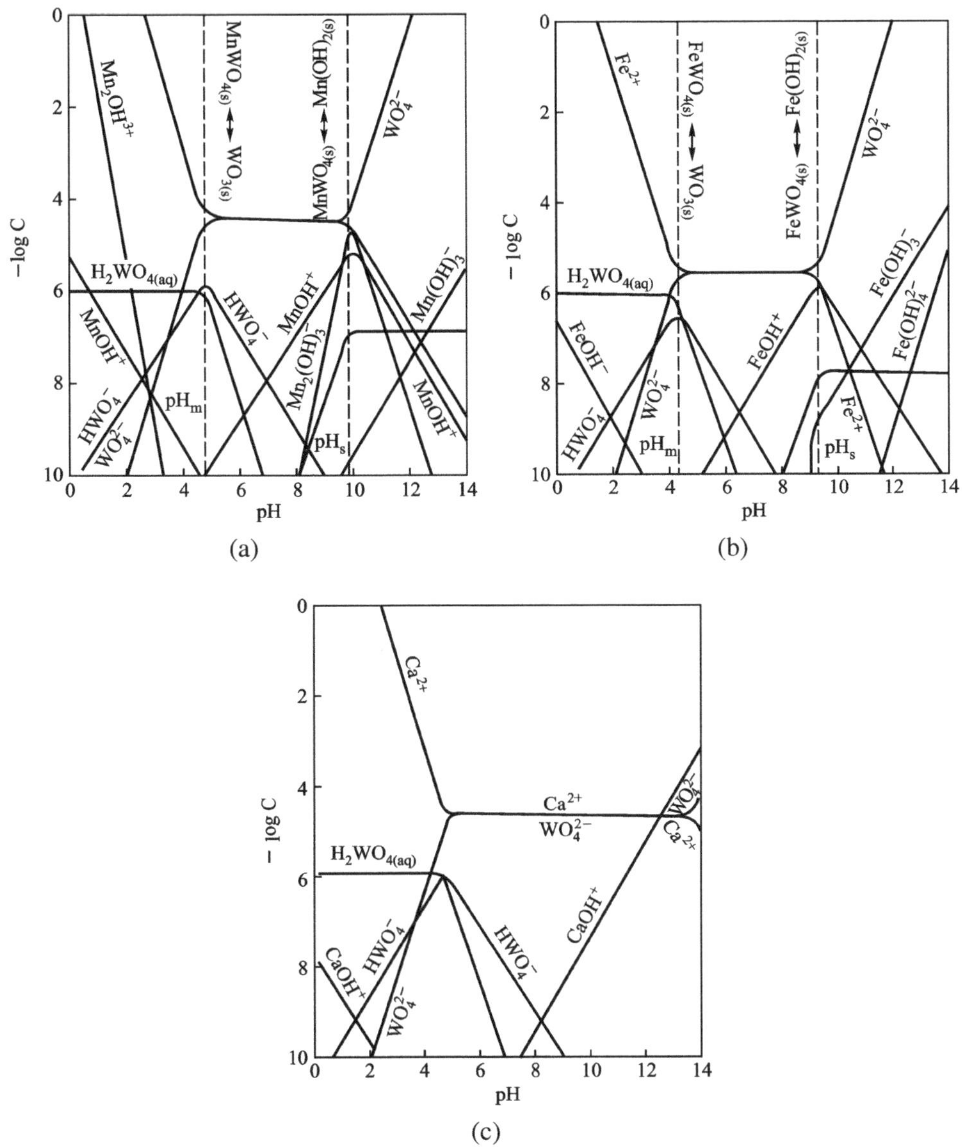

Fig. 3.6. Species distribution diagrams for mineral–water open systems: (a) hubnerite, (b) ferberite, and (c) scheelite (Hu and Wang, 1985).

surface is positive due to the presence of more $MnOH^+$ and $FeOH^+$ than HWO_4^- till both become equal around pH = 2.8 (i.e. PZC). When pH increases to 4.8, both HWO_4^- and WO_4^{2-} predominate and the surfaces exhibit a negative potential.

It is interesting to note that there is a "near PZC region" (slightly negative) where predominant species Mn^{2+}, Fe^{2+} and WO_4^{2-} become equal to each other in the pH range

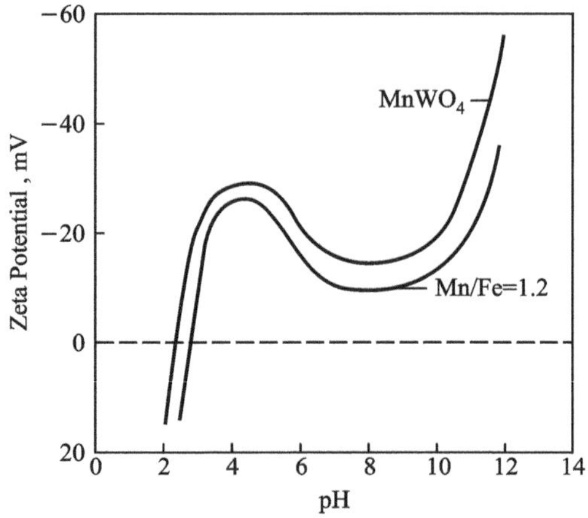

Fig. 3.7. Zeta-potential of wolframites with various Mn/Fe ratios (Hu and Wang, 1985).

between 6.0 and 9.3. Regardless of the change in the Mn/Fe ratio, the "near PZC region" remains around neutral pH. When pH is ~ 9.3, WO_4^{2-} sharply increases and makes surface more negative. Experimental results obtained (Fig. 3.7) are in good agreement with the above theoretical prediction.

It is evident from this discussion that dissolution equilibria of sparingly soluble minerals can play a major role in determining the surface properties of mineral particles. Selective hydrophobization of such particles using surfactants is the key to flotation separation.

3.4. SURFACE CONVERSION OF MINERALS BY INTERACTIONS WITH DISSOLVED MINERAL SPECIES

3.4.1. Carbonate-phosphate minerals

For the above sparingly soluble minerals, the effect of dissolved species on interfacial properties can be marked. Results obtained for the zeta-potential of apatite and calcite in water and in 2×10^{-3} M KNO_3 solutions are given in Fig. 3.7. It can be seen that the isoelectric points of calcite and apatite in both water and KNO_3 solutions are about 10.5 and 7.4, respectively. The effect of the supernatant of calcite on the zeta-potential of apatite is also shown in Fig. 3.8.

Examination of the results shows that the apatite surface is more positively charged in calcite supernatant than in water over the entire pH range investigated, with the isoelectric point shifting from 7.4 to about 11 towards that of calcite. Similarly, the zeta-potential of calcite in apatite supernatant as a function of pH, given in Fig. 3.8 shows it to be reduced drastically by the supernatant of apatite, and the isoelectric point to shift from 10.5 to 6.5. The decrease in the magnitude of the zeta-potential values in KNO_3 solutions from those in water alone could be attributed to the compression of the double layer under high ionic

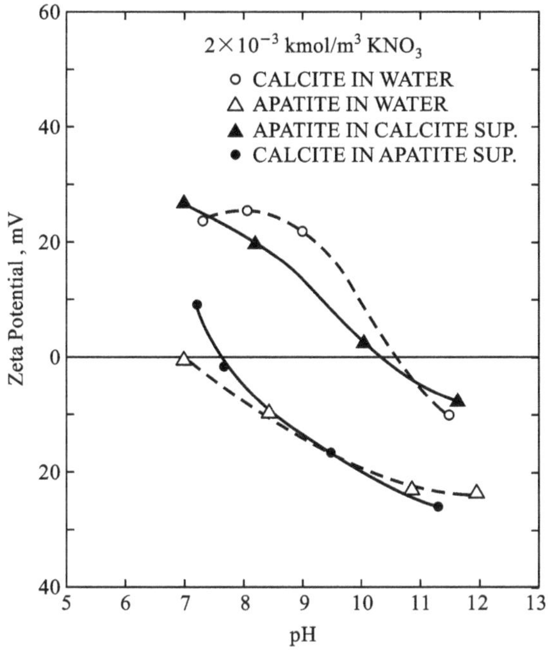

Fig. 3.8. Effect of mineral supernatants on the isoelectric points of calcite and apatite (Amankonah et al., 1985; Somasundaran et al., 1985).

strength conditions, but the shift of the isoelectric point suggests chemical changes on the surface.

In other words, these results show calcite and apatite to interchange approximately their isoelectric points in the supernatants of each other. This suggests surface conversion of apatite to calcite and calcite to apatite respectively. Implications of such conversion on selective flotation is to be noted. In flotation, the presence of flotation reagents such as oleate or other collectors actually complicates the system behavior even further and this aspect is discussed in a subsequent section.

The observed changes in the electrokinetic properties of calcite and apatite can be examined on the basis of the mineral-solution chemical equilibria involving dissolved mineral species. From studies of solubility isotherms for apatite and calcite at 25°C (Figs. 3.4 and 3.5), the singular point for these minerals is identified to be 9.3. Above this pH calcite is more stable than apatite. The implication is that, if apatite is brought in contact with alkaline solutions (pH $>$ 9.3) previously equilibrated with calcite, calcite can precipitate on the surface of apatite. In other words, if the apatite system is open to atmospheric CO_2, calcite can either precipitate from the solution and subsequently form surface coating on apatite, or the surface of apatite particles could be converted to calcite through various surface reactions leading to surface precipitation.

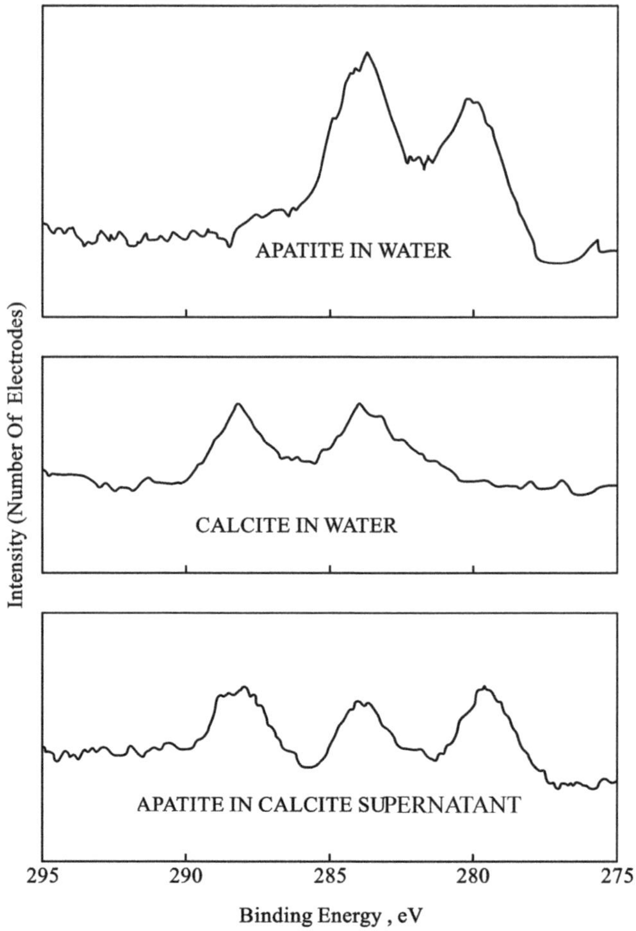

Fig. 3.9. ESCA spectra of C(1s) peak of apatite conditioned in calcite supernatant at pH 12 (Amankonah et al., 1985).

Solution conditions for the conversion of apatite to calcite by dissolved species can be more readily identified by considering the chemical reaction responsible for the process:

$$Ca_{10}(PO_4)_6(OH)_{2(s)} + 10CO_3^{2-} \xrightarrow{pH} 10CaCO_{3(s)} + 6PO_4^{3-} + 2OH^- \quad (3.38)$$

It can be seen that, depending on the pH of the solution, apatite can be converted to calcite if the total carbonate in solution is above a certain critical level. In fact, the amount of dissolved carbonate from atmospheric CO_2 does exceed that required to convert apatite to calcite under high pH conditions. Although the kinetics of such transformations is not known, such reactions can be important in determining the surface characteristics of minerals. The presence of surface converted layer on apatite or calcite has been supported by the electrokinetic results given in Fig. 3.8 and surface analysis (ESCA), Fig. 3.9

(Somasundaran et al., 1985). The ESCA results show that apatite, conditioned in calcite supernatant at pH 12 and washed, exhibits properties characteristic of calcite and apatite, suggesting partial conversion of apatite surface to calcite surface. This further supports the proposed surface conversion hypothesis.

On the other hand, most interestingly, the calcite surface can also be converted to apatite under appropriate conditions. Fig. 3.5 shows the amount of phosphate required to convert calcite to apatite (along with the total phosphate in equilibrium with apatite) far exceeds the amount required for this if the system is open to atmospheric CO_2. If the system is closed, it is seen that the conversion is possible in the entire pH range. It is interesting to note the amount of phosphate required for this process at each pH value. For example, for the closed system, only about 10^{-11} kmol/m^3 of phosphate is required to produce the conversion effect around pH 13.

Surface conversion due to reactions of the dissolved species with the mineral surface can be predicted using thermodynamic stability diagrams for heterogeneous mineral systems based on relevant mineral dissociation equilibria. This is illustrated in Fig. 3.10 for the calcite/apatite/dolomite system. The activities of Ca^{2+} species in equilibrium with various solid phases show that the singular point for calcite and apatite is 9.3. Above this pH, apatite is less stable than calcite and hence conversion of apatite surface to that of calcite can be expected in calcite–apatite system. Similarly, the calcite–dolomite and apatite–dolomite singular points occur at pH 8.2 and 8.8, respectively.

3.4.2. Sulfide minerals

Chemical equilibria for sulfide minerals have also been developed similar to the heterogeneous systems made up of calcite, magnesite, dolomite, and gypsum. These include various sulfide minerals such as covellite, millerite and pyrrhotite (Acar and Somasundaran, 1990). For example, covellite and pyrrhotite yield chemical species shown in Figs. 3.11 and 3.12. It can be seen from Fig. 3.12 that when pyrrhotite is conditioned in the covellite supernatant, surface conversion of pyrrhotite can take place. Solubility and ESCA results showed such conversion to actually take place in the mixed covellite–pyrrhotite systems.

3.5. CHARGE EQUILIBRIA OF MINERALS

3.5.1. Electrical double layer model

A significant step in delineating the physical chemistry of flotation systems was the bringing of the concept of the electrical double layer into widespread use in the interpretation of flotation phenomena (Fig. 3.13). Since the mineral particle will acquire a surface charge with respect to the aqueous phase upon immersion of it in aqueous solutions, oppositely charged species will migrate to the mineral/water interface until equilibrium conditions are established. As a result, a region of electrical inhomogeneity forms across the mineral/water interface. The excess (positive or negative) charge fixed at the mineral surface is balanced by a diffuse region of equal but opposite charge on the liquid side.

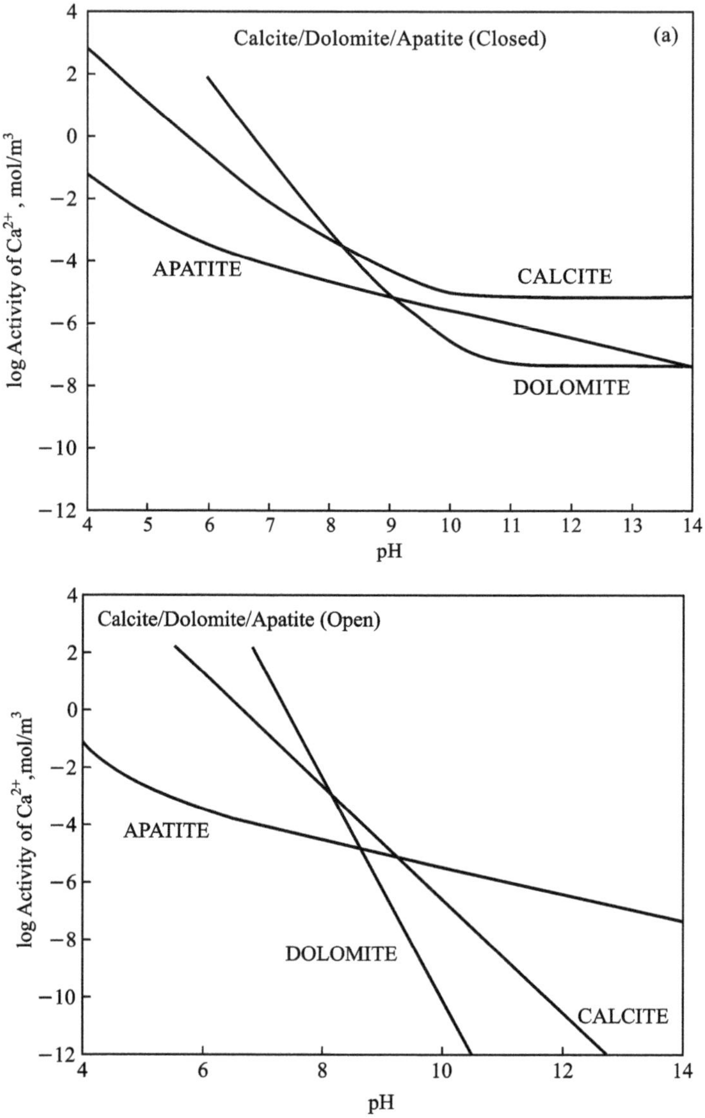

Fig. 3.10. Stability relations in calcite–apatite–dolomite systems at 25°C ([Mg] $= 5 \times 10^{-4}$ mol/l): (a) closed ([carbonate] $= 10^{-3}$ mol/l) and (b) open (Amankonah et al., 1985).

The charged surface layer and the diffuse region together constitute the diffused electrical double layer (Fuerstenau, 1976).

The electrical charge on the mineral surface can be generated by a number of mechanisms, including, preferential dissolution of surface ions, selective adsorption of ions, mineral lattice substitution or even chemisorption. The following examples illustrate how a mineral can acquire a surface charge:

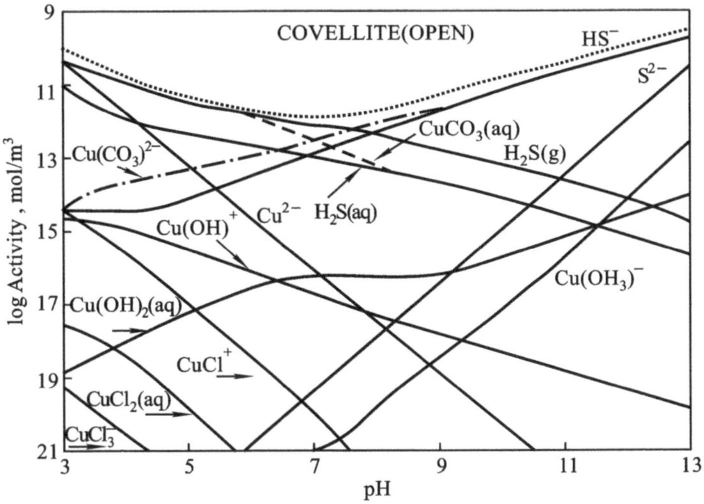

Fig. 3.11. Species distribution diagram for covellite under open (atmospheric CO_2) conditions (Acar and Somasundaran, 1990).

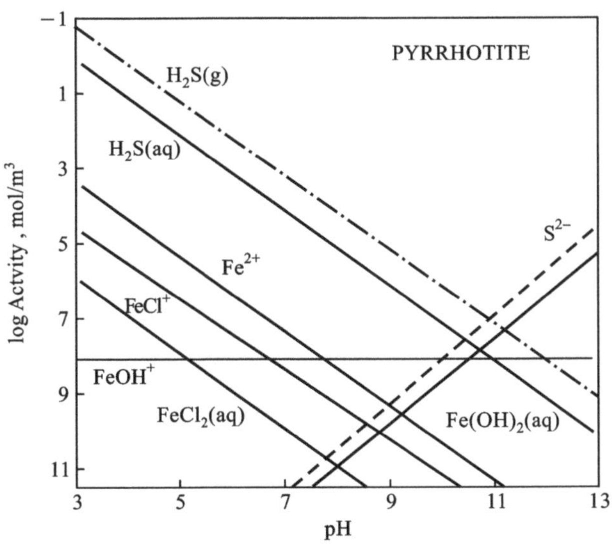

Fig. 3.12. Species distribution diagram for pyrrhotite under open conditions (Acar and Somasundaran, 1990).

(a) Transport of ions constituting the lattice across the interface (adsorption of Ag^+ and S^{2-} on Ag_2S, Ba^{2+} and SO_4^{2-} on barite).

(b) Ionization of hydroxylated surface groups (adsorption/dissociation of H and OH on oxides).

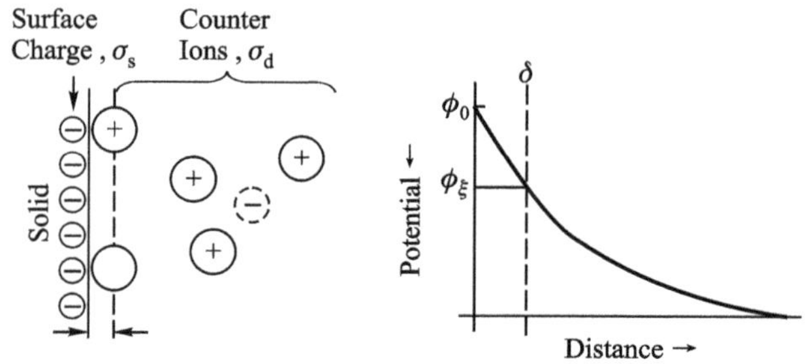

Fig. 3.13. Schematic representation of the electrical double layer and the potential drop across the double layer at the solid/water interface.

(c) Isomorphous substitution in the lattice of solids (replacement of Si^{4+} by Al^{3+} in clays and other sheet silicates).

(d) Polarization by applying a potential from an external source.

Ionic species play the predominant role in establishing the nature of the electrical double layer. Ions that are free to transfer between solid and liquid phases and thus give rise to the surface charge are called potential-determining ions, and these are specific for each mineral. In the case of the ionic solids such as $BaSO_4$, AgI and CaF_2, the surface charge arises from the preference of one of the lattice ions for sites at the solid surface as compared to the aqueous phase. Equilibrium is attained when the electrochemical potential, i.e., of these ions is constant throughout the system ($\mu = \mu_0 + vF\varphi$), where φ is the Galvani potential in the phase. Similarly for apatite $Ca_5(PO_4)_3(OH)_2$, the potential-determining ions are Ca^{2+}, PO_4^{3-} and OH^-, with the other hydrolysis products also functioning in this role because of the chemical equilibria involving all the species in this system. For oxides which constitute the most important class of nonmetallic mineral, H^+ and OH^- ions have long been considered to be potential-determining. Oxide minerals form hydroxylated surface when in contact with water vapor, or when the solid is in equilibrium with an aqueous solution. Adsorption-dissociation of H^+ and OH from the surface hydroxyls can account for the surface charge on the oxide (Yopps and Fuerstenau, 1964):

$$\mathrm{MOH_{(surf)}} = \mathrm{MO^-_{(surf)}} + \mathrm{H^+_{(aq)}}$$
$$\mathrm{MOH_{(surf)}} + \mathrm{H^+_{(aq)}} = \mathrm{MOH^+_{2(surf)}} \tag{3.39}$$

Parks and de Bruyn (1962) postulated a different mechanism for the charging of oxide surfaces, involving partial dissolution of the oxide and formation of hydroxyl complexes in solution followed by adsorption of these complexes:

$$\mathrm{M_2O_{3(solid)}} + 3\mathrm{H_2O} = 2\mathrm{M(OH)_{3(aq)}}$$
$$\mathrm{M(OH)_{3(aq)'}} = \mathrm{M(OH)^{3-m}_{m\,(aq)}} + (3-m)\mathrm{OH^-}$$
$$\mathrm{M(OH)^{3-m}_{m\,(aq)}} = \mathrm{M(OH)^{3-m}_{(surf)}} \tag{3.40}$$

The generation of surface charge by either of these mechanisms, or even by direct adsorption of H^+ and OH^-, would result in an equivalent change in the pH of the solution.

The importance of the PZC lies in the effect of the sign of the surface charge below and above it on the adsorption of all the ions and particularly those ions charged oppositely to the surface functioning as the counterions to maintain electroneutrality. In contrast to the situation in which the potential-determining ions are special for each system, any ion present in the solution can function as the counterion. If the counterions are adsorbed only by electrostatic attraction, they are called indifferent ions. Because of the thermal agitation, rather than staying attached to the surface, these ions occur in a diffuse layer that extends from the surface out into the bulk solution (Kruyt, 1952). The thickness of the diffuse double layer is $1/\kappa$, where κ is given by:

$$\kappa = (8\pi F^2 z^2 C/\varepsilon RT)^{1/2}, \tag{3.41}$$

where ε is the dielectric constant of the liquid and z is the valence of the ions. For 1–1 valent electrolyte, $1/\kappa$ is 100 nm in 10^{-5} M and 1 nm in 10^{-1} M solution, for example.

The charge in the diffuse double layer, given by the Gouy–Chapman relation (Verwey and Overbeek, 1948) as modified by Stern, is:

$$\sigma_\mathrm{d} = -\sigma_\mathrm{s} = -[(2\varepsilon RT/\pi)C]^{1/2} \sinh(zF\psi_\mathrm{d}/RT) \tag{3.42}$$

Further, if σ_d does not change appreciably, this equation shows that the adsorption density of counterions should vary as the square root of the concentration of added electrolyte, as de Bruyn (1955) has found for the adsorption of dodecylammonium acetate on quartz at low concentrations. On the other hand, some ions exhibit specific surface activity, in addition, due to electrostatic attraction and adsorb strongly in the Stern plane because of such phenomena as covalent bond formation, hydrophobic bonding, hydrogen bonding, and solvation effects. Flotation collectors generally fall into this category. Because of their specific surface activity, such counterions can adsorb in the Stern plane in amounts higher than that necessary to neutralize the surface charge and thus cause a reversal of the charge.

In many flotation systems, the electrical nature of the mineral/water interface controls the adsorption of collectors. The flotation behavior of insoluble oxide minerals, for example, is best understood in terms of electrical double-layer phenomena. A very useful tool for the study of these phenomena in mineral/water systems is the measurement of electrokinetic potential, which results from the interrelation between mechanical fluid dynamic forces and interfacial potentials. Two methods most commonly used in flotation chemistry research for evaluation of the electrokinetic potential are electrophoresis and streaming potential.

Electrophoresis involves measurement of the rate at which fine particles move under an electrical field while streaming potential technique is based on the measurement of the potential generated when a liquid is forced through a bed of the particles. Using appropriate theories, it is possible to estimate potential at the slip plane, which is considered to be just outside the Stern plane. The potential at the slipping plane is referred to as the zeta potential, ζ, and is generally assumed to approximate the Stern layer potential, ψ. This approximation seems valid because the potential difference between the Stern plane and the slipping plane is small compared with the total potential difference across the double layer. Only those ions in the diffuse layer outside the slipping plane are involved in the

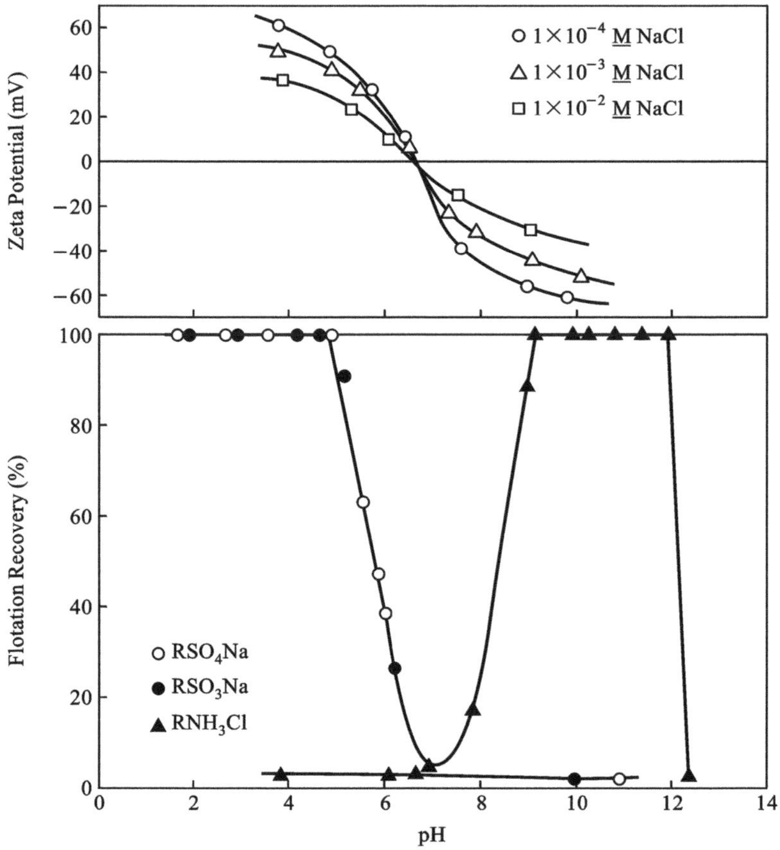

Fig. 3.14. Dependence of flotation of goethite on surface charge. Upper graph is zeta-potential as a function of pH at different concentrations of sodium chloride; lower graph is flotation recovery in 1×10^{-3} mol/l solutions of dodecylammonium chloride, sodium dodecyl-sulfate and sodium dodecyl-sulfonate (Iwasaki et al., 1960).

electrokinetic process. Besides the electrokinetic phenomena themselves, electrokinetic techniques provide a method for measuring electrical effects due to adsorption. Monitoring of the change in zeta potential as solution conditions are varied is useful in this regard since from these changes, modes of adsorption of ions involved can be ascertained. In addition to controlling the collector adsorption, the double layer governs the stability of suspensions as well as kinetics of bubble-particle contact.

PZC and IEP are important parameters for surface characterization of oxide minerals. The flotation of these minerals is best understood in terms of the electrical double layer theories. Simple oxide minerals such as hematite, goethite, magnetite and corundum float well with cationic collectors above their PZC. Fig. 3.14 shows the flotation of goethite using both anionic and cationic collectors. The PZC of this mineral is pH 6.7. Anionic collectors are effective for goethite below pH 6.7 since the mineral is then positively charged,

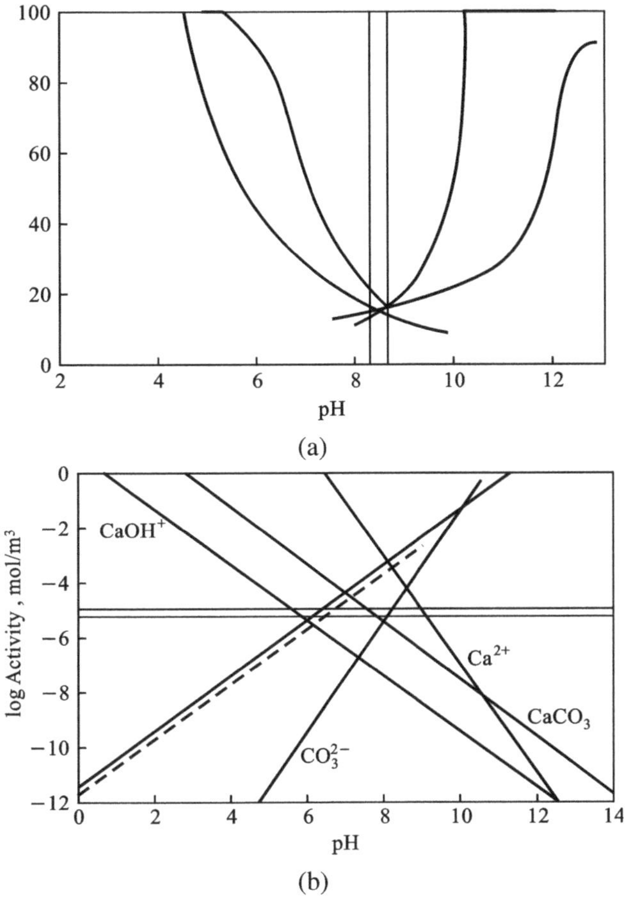

Fig. 3.15. (a) Dependence of flotation of calcite on PZC using dodecylammonium acetate (DDA) and sodium dodecylsulfate (SDS); (b) determination of PZC from species distribution diagram for calcite.

and cationic collector, DAC, work at above pH 6.7 since the surface charge is negative under those conditions.

Similar relationship can be seen in the flotation of complex minerals such as calcite (Fig. 3.15). Calcite undergoes dissolution reactions in solution and forms the following dissolved species: HCO_3^-, CO_3^{2-}, Ca^{2+}, $CaHCO_3^+$ and $CaOH^+$. From the species distribution diagram (Fig. 3.15b), the isoelectric point can be obtained assuming correspondence between it and PZC of the solution, by estimating the pH at which the total concentration of negative ions is equal to that of the positive ions. For calcite, a value of 8.2 was obtained for the PZC, in close agreement with that obtained from measurements of solution pH changes and flotation tests (Fig. 3.15a). Cationic dodecyl ammonium acetate floats the mineral at $pH > 8.2$ while dodecylsulfate works well at $pH < 8.2$ (Hanna and Somasundaran, 1976).

The above clearly shows the predominant role played by the electrochemical properties of the minerals in determining surfactant adsorption on minerals and their flotation (Somasundaran and Agar, 1967).

REFERENCES

Acar, S., Somasundaran, P., 1990. Minerals & Metallurgical Processing 7, 94–99.
Amankonah, J.O., et al., 1985. Colloids & Surfaces 15, 295–307.
Ananthapadmanabhan, K.P., Somasundaran, P., 1985. Colloids & Surfaces 13, 155–167.
Butler, J.N., 1967. Ionic Equilibrium. Adison-Wesley, Reading, MA, p. 287.
Clark, S.W., Cooke, S.R.B., 1968. Trans. AIME 241, 334.
de Bruyn, P.L., 1955. Trans. AIME 202, 291.
Fuerstenau, M.C., et al., 1976. In: Fuerstenau, M.C. (Ed.), Flotation—Gaudin Memorial Volume, vol. 1. AIME, pp. 148–196.
Garrels, R.M., Christ, C.L., 1965. Solutions, Minerals and Equlibria. Freeman, Cooper & Company, San Francisco.
Hanna, H.S., Somasundaran, P., 1976. In: Fuerstenau, M.C. (Ed.), Flotation—Gaudin Memorial Volume, vol. 1. AIME, p. 197.
Hu, Y.H., Wang, D.Z., 1985. Nonferrous Metals 37, 26–32.
Hu, Y., Wang, D., 1992. J. CSIMM 23 (3), 273–279.
Hu, Y.H., Luo, L., Qiu, G.Zh., Wang, D.Z., 1995. Transactions of Nonferrous Metals Society of China 5 (1), 26–30.
Hu, Y.H., Qiu, G.Zh., Xu, J., Wang, D.Z., 1996a. J. of Mining and Metallurgy 1, 28–33.
Hu, Y., Qiu, G., Wang, D., 1996b. Nonferrous Metals 48 (2), 40–44.
Iwasaki, I., Cooke, S.R.B., Colombo, A.F., 1960. Rep. Invest. No. 5593, U.S. Bureau of Mines.
Kruyt, H.R. (Ed.), 1952. Colloid Science, vol. 1. Elsevier, Amsterdam.
Latiner, W.M., 1952. Oxidation Potentials. Prentice-Hall, Englwood Cliffs, NJ, p. 187.
Linke, W.F., 1965. Solubilities—Inorganic and Metalorganic Compounds, vol. II. American Cynamid Co., Stamford, CT, USA.
Morgan, J.J., 1981. Aquatic Chemistry, 2nd edn. Wiley-Interscience, New York.
Nagaraj, D.R., Somasundaran, P., 1981. Trans. AIME/SME 270, 1351–1357.
Parks, G.A., de Bruyn, P.L., 1962. J. Phys. Chem. 66, 967–973.
Somasundaran, P., Agar, G.E., 1967. J. Colloid Interface Sci. 24, 433–440.
Somasundaran, P., Wang, Y.H., 1984. In: Mishra, D.H. (Ed.), Adsorption and Surface Chemistry of Hydroxyapatite. Plenum Press, NY, pp. 129–149.
Somasundaran, P., Amankonah, J.O., Ananthapadmanabhan, K.P., 1985. Colloids & Surfaces 15, 309–333.
Verwey, E.J.W., Overbeek, J.Th.G., 1948. Theory of the Stability of Lyophobic Colloids. Elsevier, Amsterdam.
Yopps, J.A., Fuerstenau, D.W., 1964. J. Colloid Sci. 19, 61.

Chapter 4

Mineral–flotation reagent equilibria

Adsorption of molecules on solids from solution is an important process controlling a variety of interfacial processes such as mineral flotation, dewatering, flocculation/dispersion, blood clotting and micellar flooding of oil wells. Adsorption of surfactants on minerals from aqueous solutions results from energetically favorable interactions between the solid adsorbent and the solute species and is often a complex process since it can be influenced by the properties of solid, solvent and solute components of the system. Several interactions such as electrostatic attraction, covalent bonding, hydrogen bonding or non-polar interactions between the adsorbate and the adsorbent species, and lateral interaction between the adsorbed species as well as their desolvation can contribute to the adsorption process.

In this chapter, the interaction forces responsible for adsorption at solid–liquid interfaces and the microstructure of the adsorbed layer that influences the flotation and flocculation processes are discussed. Surface and bulk interactions between the flotation reagents and dissolved mineral species and their solution equilibria are described using multi-pronged experimental and diagrammatic schemes. Electrochemical equilibria of mineral–flotation agent are also discussed.

4.1. ADSORPTION AND INTERACTION FORCES OF REAGENTS

Adsorption can be considered to be a process of selective partitioning of the adsorbent species to the interface in preference to the bulk and is the result of interactions of such species with the surface species on the solid. The interactions responsible for adsorption can be either physical or chemical in nature. Adsorption in solution is a more complicated phenomenon than that in gas owing to the competition for adsorption sites by the solvent species as well.

Adsorption can be broadly classified into two categories, *physical adsorption* and *chemical adsorption*, depending on the nature of the forces involved and usually referred to as physisorption and chemisorption respectively (Adamson, 1982). Physical adsorption is usually weak and reversible and involves small energy changes. The van der Waals forces and electrostatic forces are primarily responsible for physical adsorption, which is also characterized by a high rate of adsorption and formation of multilayers (Parfitt and Rochester, 1983). Chemical adsorption occurs through covalent bonding between the adsorbate and the surface species on the solid. Chemical adsorption normally involves an activation stage and is characterized by relatively higher energy changes and a lower rate of adsorption. This adsorption is usually strong and irreversible and is limited to a monolayer. A distinction between physical and chemical adsorption can usually be made from the temperature dependence of the adsorption process. In the case of physical adsorption, the adsorption generally decreases with temperature while in the case of chemisorption it

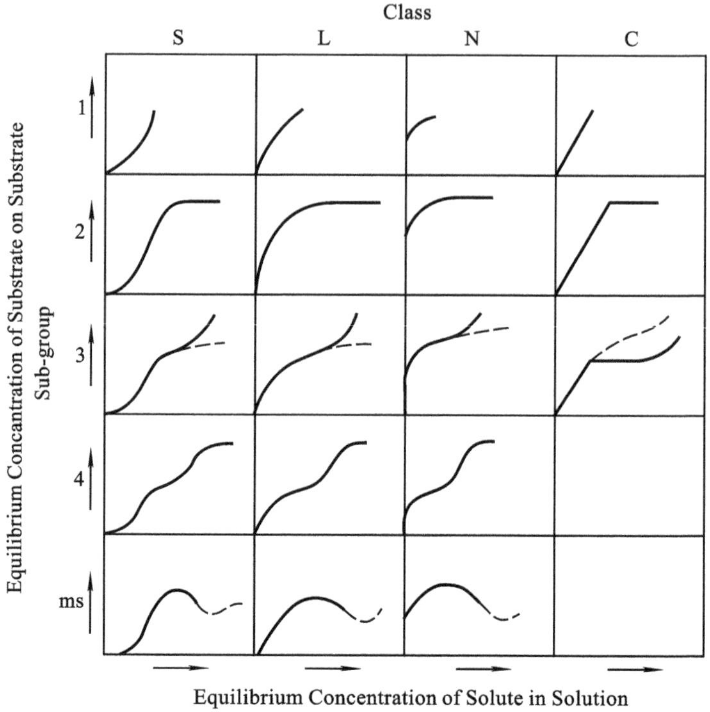

Fig. 4.1. Classification of adsorption isotherm (after Giles et al., 1960).

increases with temperature. However it must be noted that the distinction between physical and chemical adsorption is an arbitrary one and in many cases an intermediate character of adsorption is encountered. In some cases like adsorption of gases on metal surfaces, physisorption may take place initially and may be followed by adsorbent-adsorbate reactions resulting in chemisorption.

Adsorption isotherms are commonly used to describe adsorption processes and these represent a functional relationship between the amount adsorbed and the activity of the adsorbate at a constant temperature. The shape of the adsorption isotherm gives useful information regarding the mechanisms of the adsorption process. A classification of adsorption phenomena based on the shape of the isotherms is given by Giles et al. (1960) as shown in Fig. 4.1. Mainly four major classes of isotherms have been identified based on the initial part of the isotherms: (a) *S-type isotherm* with a convex shaped initial portion where adsorption rate increases with adsorption density and is indicative of vertical orientation of adsorbed molecules at the surface; (b) *L-type (Langmuir type) isotherm*, characterized by a concave initial region, represents systems in which the solvent is relatively inert and adsorption rate decreases with adsorption density. This is usually indicative of molecules adsorbed flat on the surface or ions vertically adsorbed with strong intermolecular attraction.

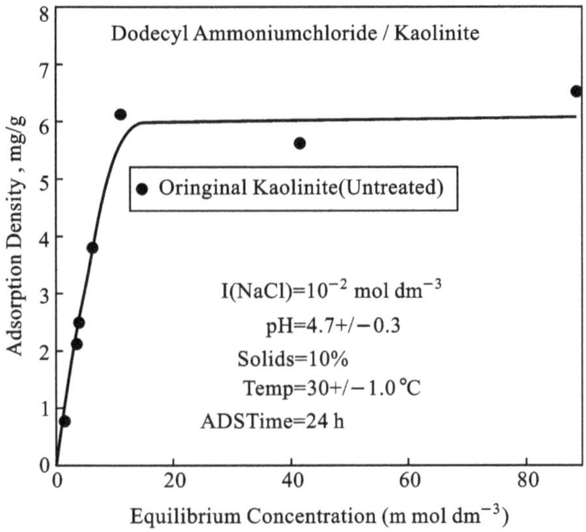

Fig. 4.2. Adsorption isotherm of dodecylammonium chloride on prewetted kaolinite.

(c) *H-type isotherm*, where the solute has such a high affinity for the solid that all of it is adsorbed at low concentrations and which is often exhibited by polymers; and (d) *C-type isotherm*, in which the initial portion is linear and indicates constant solute partitioning between the solution and the interface and this is produced by solutes which penetrate into the solid more readily than the solvent. Further classification is based on the shapes of the subsequent portions of the isotherms.

Adsorption isotherms of surface-active agents on minerals show shapes indicative of additional special mechanisms governing the adsorption process. The adsorption isotherms of dodecylammonium chloride on kaolinites are of the simple pseudo-Langmuir type (Fig. 4.2) where the adsorption increases linearly with concentration till the surface is saturated. Adsorption isotherms of sulfonates on minerals such as kaolinite from micellar solutions exhibit special features such as a maximum or a minimum around the critical micelle concentration (CMC) (Siracusua and Somasundaran, 1986). The presence of the minimum is of much interest in processes such as tertiary oil recovery using surfactant flooding since operation under such conditions would lead to minimum surfactant loss during the process. Hanna and Somasundaran (1977) have shown the existence of an adsorption maximum in their studies of Berea sandstone/Mahogany sulfonate, kaolinite/dodecylbenzenesulfonate, limestone/Mahogany sulfonate under certain conditions of salinity (Figs. 4.3–4.5). The existence of the maximum is attributed in this case to the exclusion of surfactant aggregates from the structured interfacial region owing essentially to their incompatibility with this region. Chromatographic separation of the surfactants can also lead to adsorption maximum and hysteresis during desorption (Siracusua and Somasundaran, 1986). The presence of structure breaking ions such as ammonium, as opposed to structure making ions such as sodium, can lead to the disappearance of the maximum

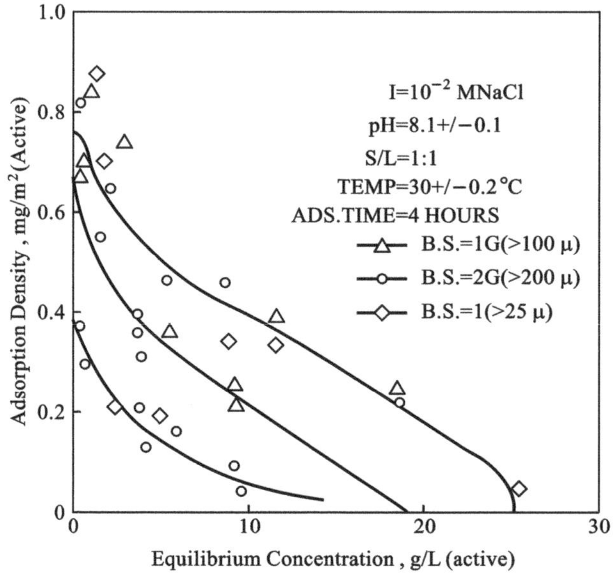

Fig. 4.3. Adsorption isotherms of Mahogany sulfonate AA on two ground Berea sandstone samples and one crushed Berea sandstone sample.

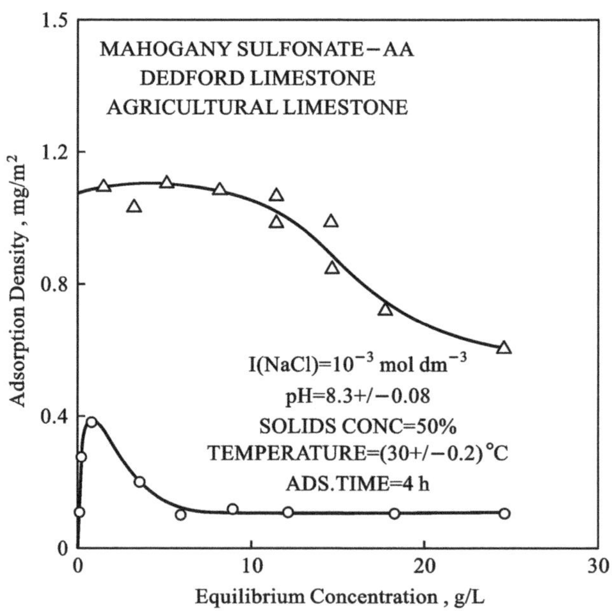

Fig. 4.4. Effects of structure making and structure breaking electrolytes on the adsorption of dodecylsulfonate on kaolinite at pH = 6.5.

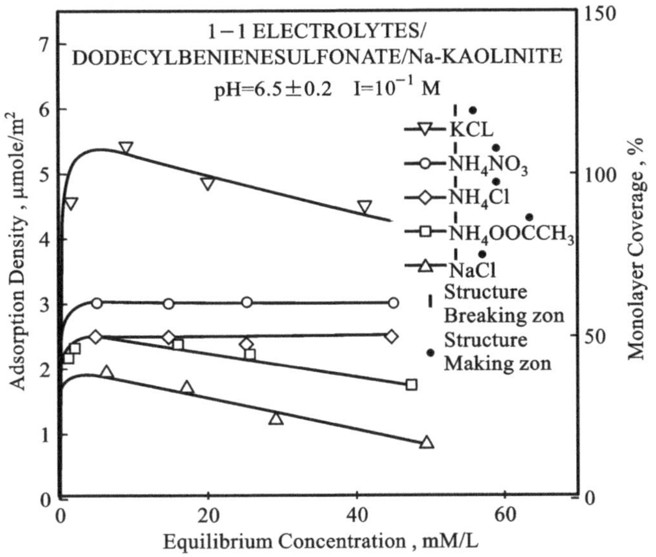

Fig. 4.5. Adsorption of Mahagony sulfonate AA on porous and non-porous limestones.

(Fig. 4.4). Mineralogical and morphological characteristics of the adsorbent can also produce significant effects on adsorption. The effect of adsorbent porosity is shown in Fig. 4.5 where the adsorption density of mahogany sulfonate is lower on porous limestone due to non-wetting of the pores and micellar exclusion. Large molecules, particularly those of the polymers can exhibit abnormal molecular weight dependence with the adsorption decreasing with increase in the molecular weight (Fig. 4.6) (Markovic and Somasundaran, 1993). This is attributed to the failure of the large molecules to enter the pores of the solid. More complex isotherm shapes are encountered as in the case of the adsorption of alkyl surfactants on silica and alumina. For example, the adsorption isotherm of sodium dodecylsulfonate on alumina consists of four regions depending on the dominant adsorption mechanism. Adsorption of polymeric reagents on minerals typically results in a pseudo-Langmuirian type isotherm as shown in Fig. 4.7 for the adsorption of polyacrylamide on Na-kaolinite (Hollander et al., 1981).

The adsorption density, which is the amount of adsorbate removed from the solution to the interface can be mathematically expressed as:

$$\Gamma_{\mathrm{i}} = 2rC\exp\left(\frac{-\Delta G^{0}_{\mathrm{ads}}}{RT}\right) \tag{4.1}$$

where Γ_{i} is the adsorption density in the plane δ, which is at the distance of closest approach of counterions to the surface, r is the effective radius of the adsorbed ions, C is the bulk concentration of the adsorbate in mol/ml, R is the gas constant, T is the absolute temperature and $\Delta G^{0}_{\mathrm{ads}}$ is the standard free energy of adsorption. In practice however the

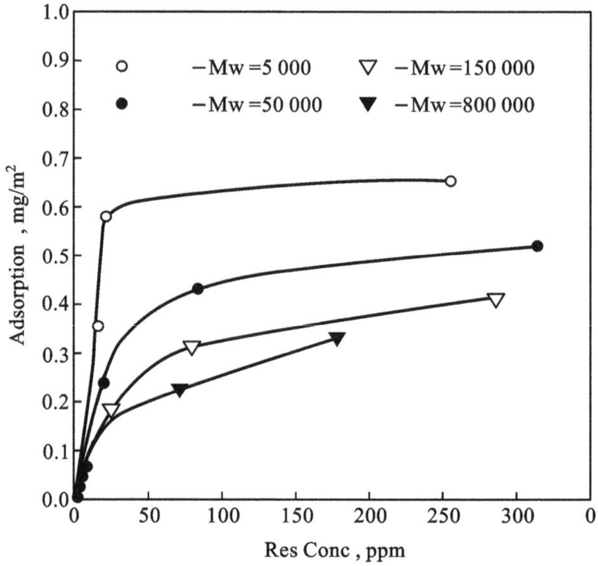

Fig. 4.6. Effect of adsorbent porosity on the adsorption of polymers of different size.

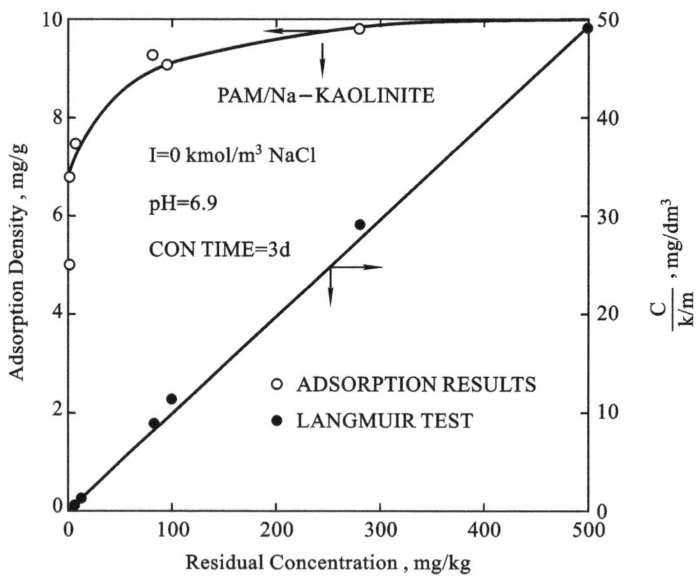

Fig. 4.7. Adsorption isotherm of polyacrylamide on Na-kaolinite as a function of residual polymer concentration and the corresponding Langmuirian plot.

adsorption density is measured as depletion of the adsorbate from the solution as shown in the following equation:

$$\Gamma = (C_{\mathrm{f}} - C_{\mathrm{i}})\frac{V}{WS}, \tag{4.2}$$

where Γ is the adsorption density in kmol/m^2, C_{f} and C_{i} are the initial and final concentrations of the adsorbate in kmol/l, V, the volume of solution in liter, W, the weight of the adsorbent in grams and S the specific surface area in m^2.

The net driving force for adsorption ΔG^0 can be considered to be the sum of a number of contributing forces.

$$\Delta G^0_{\mathrm{ads}} = \Delta G^0_{\mathrm{elec}} + \Delta G^0_{\mathrm{chem}} + \Delta G^0_{\mathrm{c\text{–}c}} + \Delta G^0_{\mathrm{c\text{–}s}} + \Delta G^0_{\mathrm{H}} + \Delta G^0_{\mathrm{H_2O}} + \cdots, \tag{4.3}$$

where $\Delta G^0_{\mathrm{elec}}$ is the electrostatic interaction term, $\Delta G^0_{\mathrm{chem}}$ is the chemical term due to covalent bonding, $\Delta G^0_{\mathrm{c\text{–}c}}$ is the lateral interaction term owing to the cohesive chain–chain interaction among the adsorbed long chain surfactant species, $\Delta G^0_{\mathrm{c\text{–}s}}$ is due to the interactions between the hydrocarbon chains and the hydrophobic sites on the solid, ΔG^0_{H} is the hydrogen bonding term, and ΔG^0 is the solvation or desolvation of the adsorbate or any other species during adsorption. For each surfactant–solid system, several of the above terms can be significant depending on the mineral and the surfactant type, surfactant concentration, background electrolyte and solvent temperature. For non-metallic minerals, electrostatic attraction and lateral interaction effects are considered to be the major factors determining adsorption while for salt-type minerals such as calcite and sulfides such as galena, the chemical term often becomes significant.

4.1.1. Electrostatic forces

Particulates are often charged in solutions either due to the preferential dissolution of surface species, hydrolysis and ionization of surface species or adsorption of various charged ions and complexes. Adsorption of oppositely charged species on such particles can take place by electrostatic interaction with the free energy of adsorption given by:

$$\Delta G^0_{\mathrm{elec}} = zF\psi_\delta, \tag{4.4}$$

where z is the valency of the adsorbate species, F is the Faraday constant and ψ_δ is the potential at the δ plane.

The role of electrostatic forces in adsorption is clearly illustrated in Fig. 4.8 where it is seen that the negatively charged anionic sulfonate adsorbs in significant amounts only below the pH corresponding to the point of zero charge (PZC $=$ 9.1) where alumina is positive (Somasundaran and Fuerstenau, 1966). Similar correlations have been obtained for adsorption of sulfonates and amines on calcite (Somasundaran and Agar, 1967). In systems where electrostatic forces play a dominant role, the presence in solution of other species charged similarly to the adsorbate species can influence adsorption owing to the competition for adsorption sites. For example, addition of potassium nitrate depresses the flotation of quartz using cationic dodecylammonium ions owing to the reduction of adsorption of amine cations by the competing potassium ions (Somasundaran, 1974). This

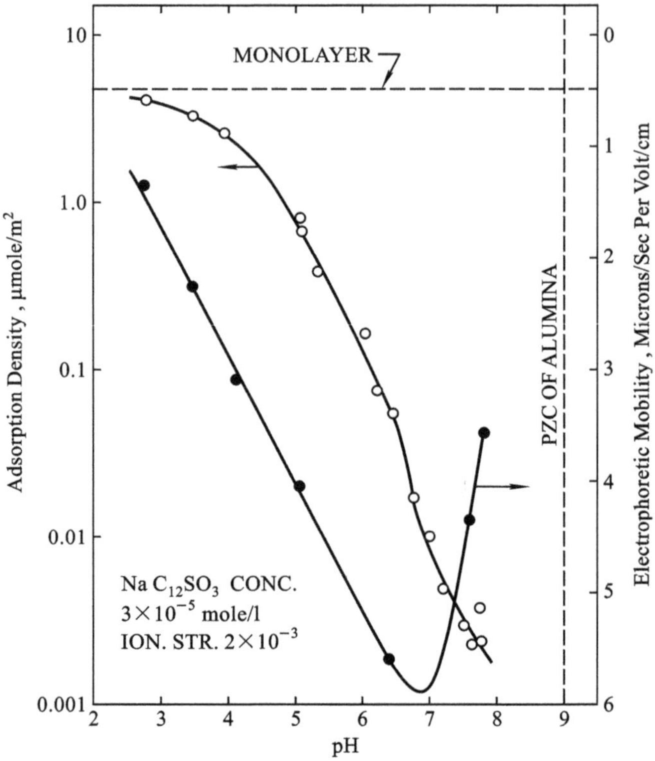

Fig. 4.8. Adsorption of dodecylsulfonate on alumina as a function of pH.

can also be considered to be due to a compression of the double layer by the potassium ions thus reducing the electrostatic potential responsible for the adsorption of dodecylammonium ions. If the added salt contains multivalent ions, this effect can be much more significant, particularly when charge reversal of the surface is caused by them. In such cases, adsorption of surfactant ions charged similarly to the surface can now take place causing activation of flotation by such ions (Fuerstenau et al., 1964; Healy et al., 1957).

In some cases, it is observed that the adsorption isotherm undergoes a sharp increase in slope at a particular surfactant concentration and this is attributed to lateral chain interactions between the adsorbed surfactant molecules (Somasundaran and Fuerstenau, 1966). A detailed thermodynamic treatment of this process is included in the next section due to the dominant role played by it in flotation systems.

4.1.2. Chemical forces

Infrared results obtained for adsorbed surfactant layers on certain minerals show that fatty acids and sulfonates can adsorb also through covalent bonding (Robert et al., 1954; Peck and Wardsworth, 1965). However identification of a bonding type on infrared spectrum does not necessarily suggest the existence of such bonding on minerals in surfactant solutions because of the possible alterations in the chemical state of the adsorbed surfac-

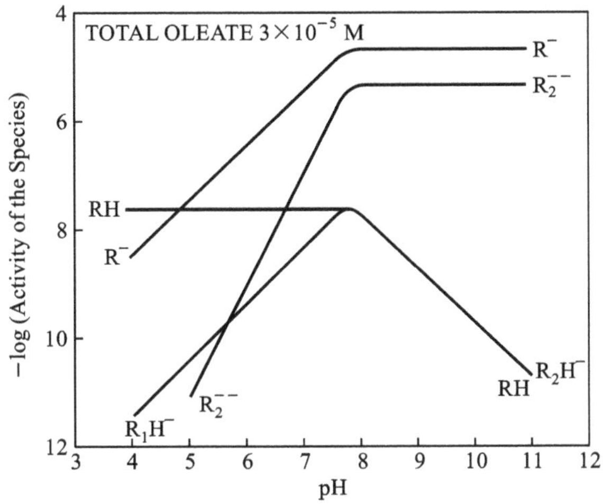

Fig. 4.9. Oleate species distribution diagram as a function of pH. Total oleate concentration = 3×10^{-5} M.

tants during preparation for spectroscopic investigations whether under dry, compressed, wet or thin film states. Chemisorption can also take place through ion-exchange. For example, in the case of adsorption of oleate on fluorite, the surfactant ions are found to release an equivalent amount of fluoride ions to form a surface layer of calcium oleate (Bahr et al., 1969; Bisling).

In mineral–reagent systems, surface precipitation has been proposed as another mechanism for chemisorption. The solubility product for precipitation and the activities of the species at the solid–liquid interface determine the surface precipitation process. Under appropriate electrochemical conditions, the activity of certain species can be higher in the interfacial region than that in the bulk solution and such a redistribution can lead to many reactions. For example, the sharp increase in adsorption of the calcium species on silica around pH 11 has been shown to be due to surface precipitation (Somasundaran and Ananthapadmanabhan, 1985; Xiao, 1990). Similar correlations have been obtained for cobalt–silica, alumina–dodecylsulfonate, calcite/apatite/dolomite–fatty acid, francolite–oleate and tenorite–salicylaldoxime systems. The chemical state of the surfactant in the solution can also affect adsorption (Somasundaran and Ananthapadmanabhan, 1985).

Various associative interactions of hydrolyzable surfactants in aqueous media can play a significant role in determining the adsorption behavior of these surfactants. For example, existence of ionomolecular complexes has been shown to have a significant effect on the adsorption of oleic acid on hematite as indicated by the flotation results (Xiao, 1990). Evidence for high surface activity of mixed acid-soap was obtained by surface tension measurements of oleate solutions (Ananthapadmanabhan, 1980). The surface activity of acid-soap was estimated to be larger than that of both the corresponding acid molecule and ionic soap. Similarly maximum flotation of quartz with alkylamine observed around pH 10.2 has been proposed to involve the formation of ion/dipole complexes (Somasundaran,

1976). Correlation of oleate adsorption and flotation maximum at about pH 7.5 for a variety of minerals and high abstraction (adsorption + surface precipitation) below this pH with the species distribution diagram (Fig. 4.9) suggests that the role of acid-soap dimer and precipitated oleic acid can be significant in controlling the adsorption and resultant flotation behavior.

4.1.3. Other forces

Other factors that play a major role in certain systems include hydrogen bonding, hydrophobic bonding and solvation or desolvation of species. Hydrogen bonding can be strong particularly for the adsorption of surfactants containing phenolic groups (Parks, 1975) and possibly those containing hydroxyl, carboxylic and amine groups and polymers (Hollander et al., 1981). Due to the predominance of hydrogen bonding, charged polymers can often adsorb even on similarly charged solids in significant amounts.

Hydrophobic bonding occurs when the combination of the tendency of the hydrophobic groups of the surfactant to escape from an aqueous environment and their mutual attraction becomes large enough to permit their adsorption onto a solid adsorbent by aggregation of their chains (Somasundaran and Fuerstenau, 1966). Hydrophobic bonding occurs also on solid surfaces that are at least partially hydrophobic. In this case, the surfactant molecules may be adsorbing flat on the hydrophobic surface. Such adsorption can take place also on solids that are originally hydrophilic but have acquired some hydrophobicity owing to reactions with organic species in solution.

4.2. MICROSTRUCTURE OF ADSORBED LAYER OF SURFACTANTS AND POLYMERS ON MINERALS

An isotherm adsorption of charged surfactants on oxides is shown in Fig. 4.10. This isotherm is characterized by four regions, attributed to four different dominant mechanisms being operative in each region. Mechanisms involved in these regions may be viewed as follows:

- Region I, which has a slope of unity under constant ionic strength conditions, is characterized by the existence of electrostatic interactions between the ionic surfactant and the oppositely charged solid surface.
- Region II is marked by a conspicuous increase in adsorption which is attributed to the onset of surfactant aggregation at the surface through lateral interaction between hydrocarbon chains. This phenomenon is referred to as hemi-micelle formation (Gaudin and Fuerstenau, 1955; Somasundaran and Fuerstenau, 1966). Such colloidal aggregates formed on the surface are referred to as *solloids* (surface colloids) (Somasundaran and Kunjappu, 1989) in general since many surface aggregates may have forms other than that characterized by half-micelles. In the case of simple ionic surfactants, such aggregates have also been called in addition to hemi-micelles, admicelles and surfactant self assemblies (Harwell and Bitting, 1987).

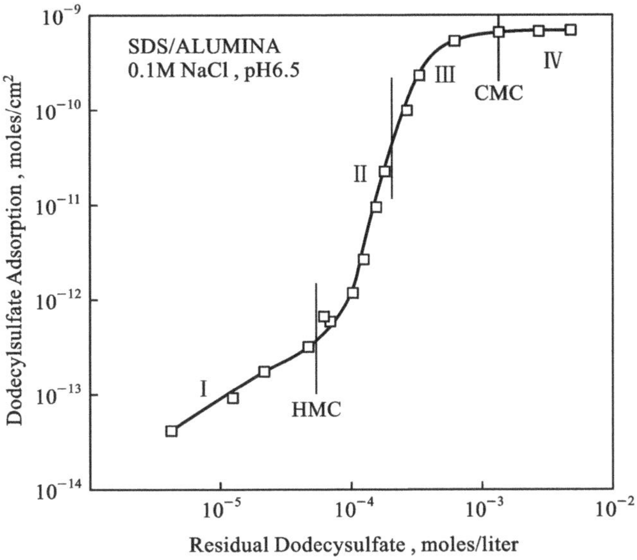

Fig. 4.10. Adsorption isotherm of sodium dodecyl sulfate (SDS) on alumina at pH 6.5 in 10^{-1} kmol/m^3 NaCl.

- Region III shows a measurable decrease in the slope of the isotherm and this is ascribed to the increasing electrostatic hindrance to surfactant adsorption following interfacial charge reversal caused by the adsorption of the charged species in this region on an adsorbate that is now similarly charged.
- Region IV and the plateau in it correspond to the maximum surface coverage as determined by micelle formation in the bulk or monolayer coverage whichever is attained at the lowest surfactant concentration; further increase in the surfactant concentration does not alter the adsorption density. A schematic representation of adsorption by lateral interactions is given in Fig. 4.11.

The surfactant aggregation process on the solid surface as described above has been widely accepted to account for the characteristic adsorption behavior of long chain surfactants on many solids. Marked changes in a number of interfacial properties such as flotation, suspension stability and contact angle have provided supporting evidence for lateral associative interaction between the hydrocarbon chains of the adsorbed surfactant species. Such an association in the interfacial region is analogous to micellar formation in the bulk solution which results from favorable energetics of partial removal of the alkyl chains from the aqueous environment. The free energy of solloid formation can be calculated in a number of ways (Somasundaran and Lin, 1971; Somasundaran et al., 1964). For example, the data obtained for the zeta potential of quartz as a function of the concentration of amines with different chain lengths (Fig. 4.12) can be treated using the following

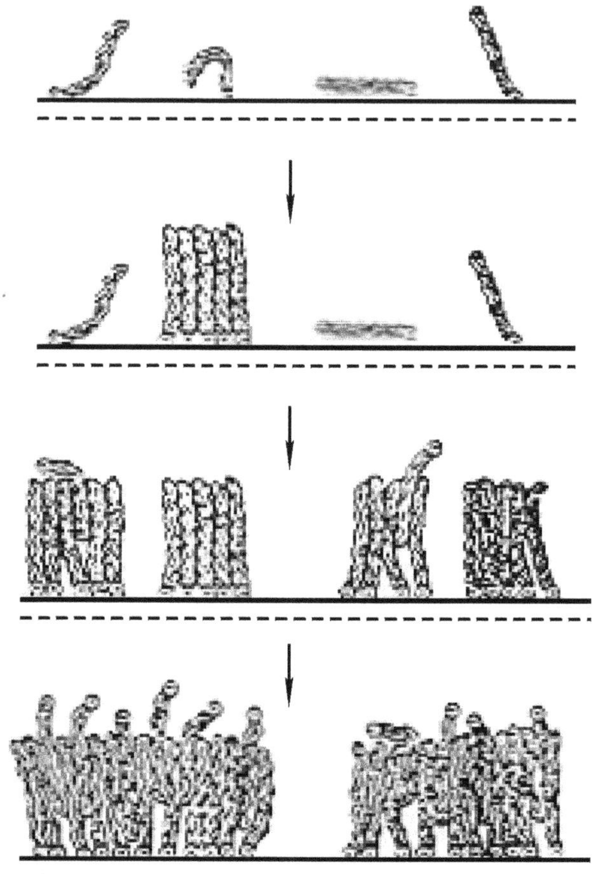

Fig. 4.11. Schematic representation of the correlation of surface charge and the growth of aggregates for various regions of the adsorption isotherm depicted in Fig. 4.10.

adsorption equation:

$$\Gamma_i = C_b \exp\left(\frac{-\Delta G^0_{elec} - \Delta G^0_{cc}}{RT}\right), \tag{4.5}$$

where ΔG^0_{ads} is divided into two terms, one corresponding to the electrostatic forces, ΔG^0_{el}, and other corresponding to the hydrocarbon chain association during solloid formation, ΔG^0_{cc}.

Concentrations corresponding to solloid formation (C_{csc}) were obtained from Fig. 4.12 by graphically locating the concentrations at which the extrapolation of the straight line portion of the zeta potential curves intersect the curve for ammonium acetate. Substitution of C_{csc} for C_b and $n\phi_s$ for ΔG^0_{cc}, where n is the number of $-CH_2-$ and $-CH_3$ groups in the chain and ϕ_s the average free energy transfer of a mole of $-CH_2-$ groups from the aqueous

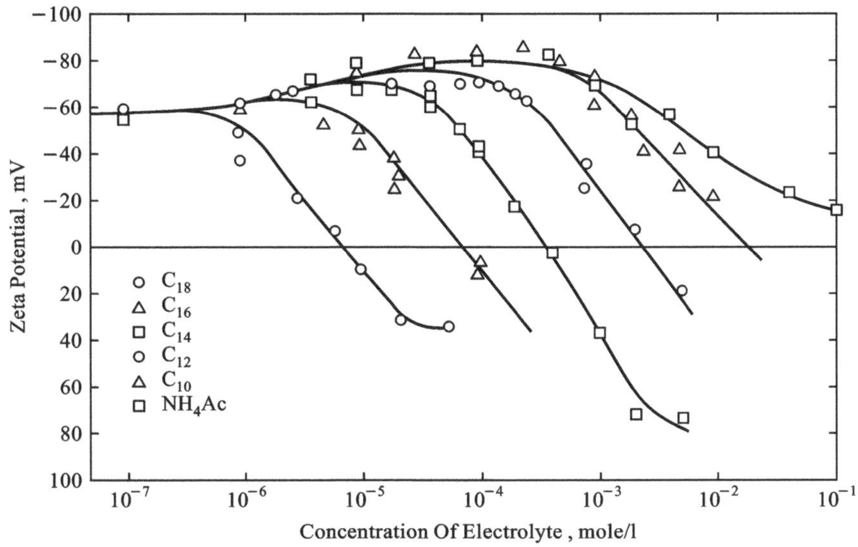

Fig. 4.12. Effect of the hydrocarbon chain length on the zeta potential of quartz in solutions of alkyl-ammonium acetates.

environment into the solloids, yields

$$\ln C_{\mathrm{csc}} = \frac{n\phi_{\mathrm{s}}}{RT} + \Delta G^0_{\mathrm{el}} \ln \frac{\Gamma_{\mathrm{i}}}{\tau} \tag{4.6}$$

A least square plot of $\log C_{\mathrm{csc}}$ as a function of number of carbon atoms in the alkyl chain is given in Fig. 4.13. ϕ_{s} obtained from the slope of this line is $-0.95kT$ and is comparable to the free energy of micellization measured for similar surfactants in solution (Fig. 4.14). Heat and standard entropy changes associated with the adsorption process can be calculated by considering the adsorption of the long-chain molecules, X, as follows (Somasundaran and Fuerstenau, 1972).

Consider the process of transfer of a solute from the solution to the surface in two steps:

1.

$$\mathrm{X(soln.,}\ a = 1) \rightarrow \mathrm{X(soln.,}\ a = C_1) \qquad \Delta G^0_{T_1} = RT_1 \ln C_1 \tag{4.7}$$

2.

$$\mathrm{X(soln.,}\ a = C_1) \rightleftarrows \mathrm{X(surf.,}\ \Gamma), \qquad \Delta G = 0$$

$$\mathrm{X(soln.,}\ a = 1) \rightarrow \mathrm{X(surf.,}\ \Gamma) \qquad (\Delta G^0_{\mathrm{x}_\sigma})_{T_2} = RT_1 \ln C_1 \tag{4.8}$$

where C_1 is the concentration of the long chain surfactant in solution in mole per liter corresponding to its adsorption density Γ in mole per sq cm at temperature T_1 and $(\Delta G^0_{\mathrm{x}})_{T_1}$ relative partial molar free energy of the adsorbed ions at this temperature. A solute standard state is considered for X in the solution such that the activity coefficient $\gamma_{\mathrm{x}} \rightarrow 1$ as $\mathrm{C} \rightarrow 0$.

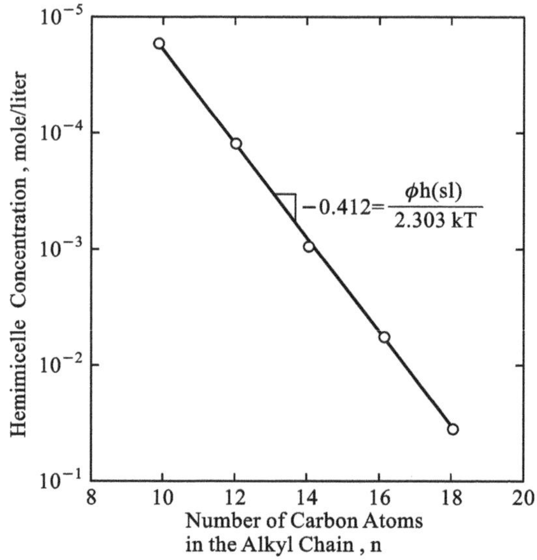

Fig. 4.13. Least square plot of C_{csc} of alkyl ammonium acetates from zeta potential data as a function of the number of carbon atoms in the chain.

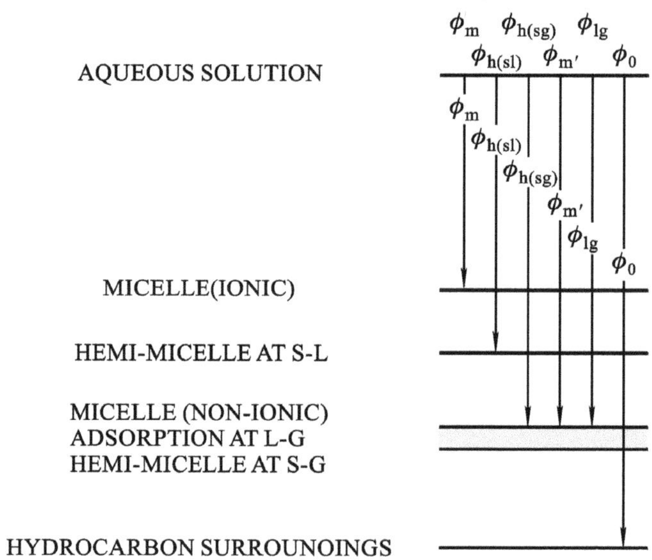

Fig. 4.14. Schematic diagram for free energy of transfer of CH_2 groups from aqueous solution to various environments [I.J. Lin and P. Somasundaran, JCIS].

Similarly at temperature T_2,

$$\left(\Delta G^0_{x_\sigma}\right)_{T_2} = RT_2 \ln C_2 \tag{4.9}$$

Knowing C_1 and C_2 for the same adsorption density, one can evaluate $\Delta G^0_{x_\sigma}$ at both temperatures from which the heat and entropy terms involved can be calculated.

$\Delta H^0_{x_\sigma}$ and $\Delta S^0_{x_\sigma}$ are the changes in heat content and entropy due to adsorption as well as dilution to the bulk concentration under consideration. Assuming ideal mixing and hence neglecting the heat of dilution:

$$\Delta \overline{H}_{x_\sigma} = \overline{H}_{x_\sigma} - \overline{H}^0_{x_1} \approx \Delta \overline{H}_{x_{ad}}, \tag{4.10}$$

where the right hand side is equal to the change in partial molar heat content of the alkyl ion on adsorption.

$$\Delta \overline{S}^0_{x_\sigma} = \Delta S_{dil} + \Delta \overline{S}^0_{x_{ad}} = R \ln C + \overline{S}_{x_\sigma} - S^0_{x_1} \tag{4.11}$$

$$\Delta \overline{H}^0_{x_{ad}} = R\frac{\ln C_2 - \ln C_1}{1/T_2 - 1/T_1} \tag{4.12}$$

$$\Delta \overline{S}_{x_\sigma} = R\frac{(T_1 \ln C_1 - T_2 \ln C_2)}{T_2 - T_1} \tag{4.13}$$

The corresponding standard heat and entropy can be obtained using the following equation, which includes a contribution from the entropy of dilution.

$$\Delta \overline{S}^0_{x_{ad}} = R\frac{(T_1 \ln C_1 - T_2 \ln C_2)}{T_2 - T_1} - \Delta S_{dil} \tag{4.14}$$

By evaluating ΔH^0_{el} and ΔS^0_{el} as a function of concentration, it is possible to obtain the changes in them due to such processes as solloid formation. However, in real adsorption systems complications arise due to interaction of other components. For example, when sulfonate ions adsorb on alumina, they will be displacing anions of the supporting electrolytes and perhaps water molecules bound to the surface. The measured changes in thermodynamic quantities will include those due to these effects also and will hence be different from the changes corresponding to the adsorbed sulfonate ions only.

It is also possible to evaluate the thermodynamic quantities associated with solloid formation by approximating it to a phase change and using the Clausius–Clapeyron type equation:

$$\frac{d \ln C_{csc}}{dT} = -\frac{\Delta H_s}{RT^2}, \tag{4.15}$$

where C_{csc} is the critical solloid concentration and ΔH_s is the heat of solloid formation at temperature T. The choice of the latter has some advantages in interpreting solloid formation as a two dimensional association process.

The heat and entropy of adsorption calculated in the manner outlined above for dodecyl sulfonate adsorption on alumina show marked changes at particular concentrations and are in agreement with the hypothesis of lateral interaction of surfactants to form two-dimensional aggregates (Fig. 4.15). Most interestingly, the association was found to produce a net increase in entropy of the system, suggesting a decrease upon aggregation

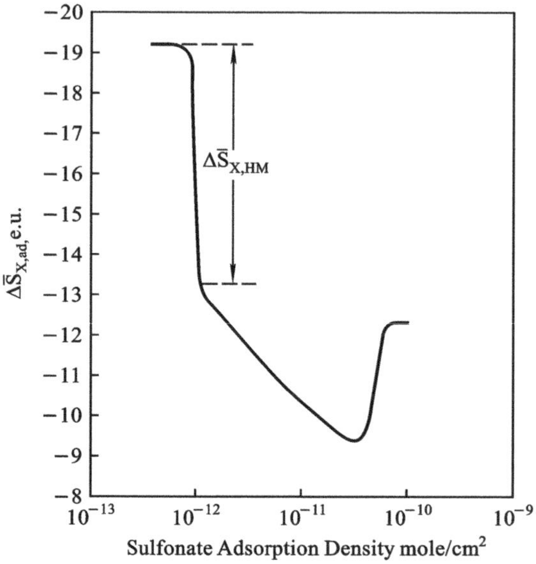

Fig. 4.15. Partial molar entropy of dodecyl sulfonate as a function of its adsorption density.

in the ordering of water molecules that were originally surrounding the isolated surfactant chains (Fig. 4.15). The concentration at which the aggregation begins depends on the pH and hence the surface potential of the particles, temperature and the chemical state and structure of the surfactant species.

4.2.1. Spectroscopic probing

Understanding of the structure of the adsorbed surfactant and polymer layers at a molecular level is helpful for improving various interfacial processes by manipulating the adsorbed layers for optimum configurational characteristics. Until recently, methods of surface characterization were limited to the measurement of macroscopic properties like adsorption density, zeta-potential and wettability. Such studies, while being helpful to provide an insight into the mechanisms, could not yield any direct information on the nanoscopic characteristics of the adsorbed species. Recently, a number of spectroscopic techniques such as fluorescence, electron spin resonance, infrared and Raman have been successfully applied to probe the microstructure of the adsorbed layers of surfactants and polymers at mineral–solution interfaces.

4.2.1.1. *Fluorescence spectroscopy*

Fluorescence emission is the radiative emission of light by an excited molecule returning to its ground state energy level. This phenomenon bears a wealth of information on the environment around the light absorbing species. Parameters of fundamental importance in fluorescence emission are: (1) emission maximum (wavelength of maximum intensity),

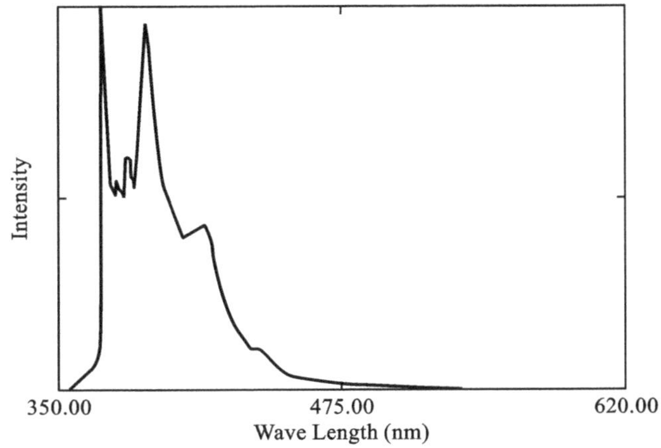

Fig. 4.16. Typical fluorescence spectrum of pyrene showing the five vibrational lines.

(2) quantum yield of fluorescence (emission efficiency measured as intensity) and (3) fluorescence life-time (time taken by the excited state to decay to $1/e$ of its initial value).

The fluorescence measurements are generally carried out by a steady state fluorescence spectrometer and life-time of fluorescence by a time resolved fluorescence life-time instrument (Lakowicz, 1983). The dependence of fluorescence intensity and life-time on the physico-chemical environment of the fluorescent probe molecule, has been well documented (Thomas). Such data have been applied to micellar photochemistry to understand the property of micelles (Turro et al., 1985). This technique has recently been adapted as a tool to probe the adsorbed layer of surfactants on solids to obtain information on the *micropolarity and microviscosity of the probe environment and the aggregation number of the surfactant species at the interface*. To determine the micropolarity, a fluorescent molecule like pyrene is used as a probe, which possesses a highly structured fluorescence spectrum with its vibrational lines susceptible to intensity fluctuations brought on by polarity changes of the medium. A properly resolved fluorescence spectrum of pyrene in fluids has five vibrational peaks in the region from 370 to 400 nm (Fig. 4.16). The intensities of the first (I_1) and the third (I_3) peaks are found to be particularly sensitive to changes in the probe environment. The ratio of these peaks (I_3/I_1), referred to as the polarity parameter, changes from about 0.6 in water to a value greater than 1.0 in hydrocarbon media.

I_3/I_1 values for pyrene were determined for alumina/sodium dodecyl-sulfate/water system for various regions of the adsorption isotherm. The data shown in Fig. 4.17 is marked by an abrupt change in local polarity of the probe from an aqueous environment to a relatively non-polar micelle type environment. This abrupt change occurs in a region that is well below the CMC and approximately coincides with the transition in the adsorption isotherm from Region I to Region II. In the plateau region, I_3/I_1 value coincides with the maximum value for SDS micellar solutions indicating the completion of aggregation on the surface. It can also be seen that the polarity parameter is fairly constant throughout most of Region III and above and hence seems to be independent of surface coverage.

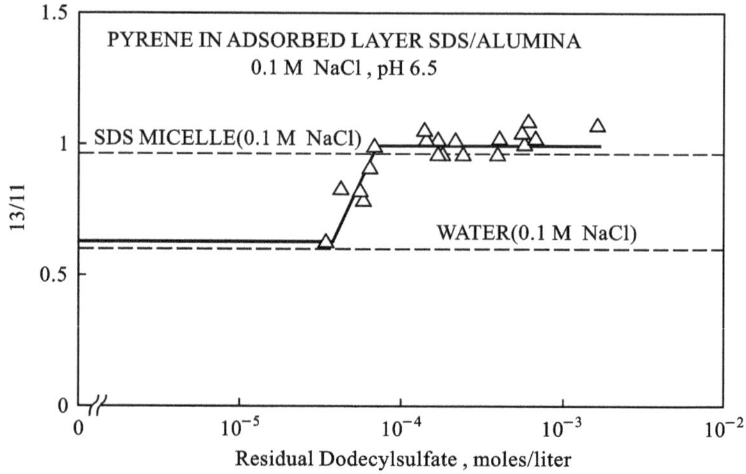

Fig. 4.17. I_3/I_1 fluorescence parameter of pyrene in sodium dodecylsulfate (SDS) absorbed layer in alumina slurries.

Information on microviscosity is obtained by studying the excimer forming capabilities of suitable fluorescent probes. The excimer, which is a complex of a ground state and excited state monomer, has a characteristic emission frequency. The intramolecular excimer formation for example, of 1,3-dinaphthyl propane (DNP), is a sensitive function of the microviscosity of its neighborhood. This property, expressed as the ratio of the excimer and monomer yield (I_e/I_m) for DNP, has been determined for dodecyl sulfonate solutions and its adsorbed layer for the various regions of the adsorption isotherm (Fig. 4.18) (Somasundaran et al., 1986). Comparing the ratios thus obtained to the I_e/I_m values of DNP in mixtures of ethanol and glycerol of known viscosities, a microviscosity value of 90 to 120 cPs is obtained for the adsorbed layer in contrast to a value of 8 cPs for micelles. The constancy of microviscosity as reported by DNP is indicative of the existence of a condensed surfactant assembly (solloids) that holds the probe.

The dynamics of fluorescence emission of pyrene has been previously studied in homogeneous and micellar solutions using time-resolved fluorescence spectrometry (Demas, 1983). While the decay kinetics of monomer and excimer emission may be derived directly for a homogeneous solution (continuous medium), statistical methods are applied to arrive at similar kinetics in aggregated micellar ensembles. This stems from a need to recognize the possibility of random multiple occupancy of the probe aggregates which affects the excimer forming probability within the aggregate. If the micellar system is viewed as groups of individual micelles with n probes, P_n the average number of probes per micelle, may be related to n, by Poisson statistics through the relation:

$$P_n = n^n \exp(-n)/n! \tag{4.16}$$

This model yields the following relation for the time dependence of monomer emission:

$$I_{m[t]} = I_{m[0]} \exp[-k_0 t + n(\exp(-k_e t) - 1)], \tag{4.17}$$

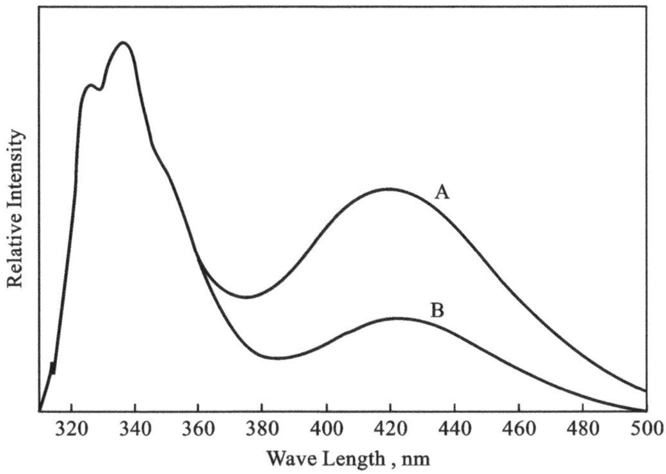

Fig. 4.18. (A) Dinaphthyl propane (DNP) fluorescence spectra in sodium-dodecyl sulfate (SDS) micellar solution and (B) SDS-alumina slurry; [SDS]/[DNP] refers to ratio of micellized or adsorbed SDS to added DNP.

where k_0 is the reciprocal lifetime of excited pyrene in the absence of excimer formation, k_e is the intra micellar encounter frequency of pyrene in excited and ground states, and $I_{m[0]}$ and $I_{m[t]}$ represent the intensity of monomer emissions at zero time and time t, respectively.

Knowing n, one can calculate aggregation number N using the expression:

$$n = [P]/[Agg] = [P]N/([S] - [S_{eq}]), \tag{4.18}$$

where [P] is the total pyrene concentration, [Agg] is the concentration of the aggregates and $[S] - [S_{eq}]$ is the concentration of the adsorbed surfactant. A kinetic analysis based on this relation was carried out from the decay profiles of pyrene in the adsorbed layer for different regions of the alumina/dodecylsulfonate adsorption isotherm (Chandar et al., 1987). The aggregation numbers for dodecyl sulfonate solloids thus obtained are marked on the adsorption isotherms in Fig. 4.19. These results yield a picture of the evolution of the adsorbed layer. The aggregates in Region II appear to be of relatively uniform size while in Region III there is a marked growth in the aggregate size. In Region II, the surface is not fully covered and enough positive sites remain as adsorption sites. Since the aggregation number is fairly constant in Region II, further adsorption in this region can be considered to occur by the formation of more aggregates. The transition from Region II to III corresponds to the isoelectric point (IEP) of the mineral, and adsorption in Region III is proposed to occur through the growth of existing aggregates rather than the formation of new ones due to lack of positive adsorption sites. Such adsorption is possible by the hydrophobic association between the hydrocarbon tails of the already adsorbed surfactant molecule and the adsorbing ones. Additional molecules adsorbing at the solid–liquid interface can be expected to orient with the ionic head turned towards the water since the solid particles posses a net negative charge under these conditions. The whole process of adsorption is portrayed schematically in Fig. 4.11.

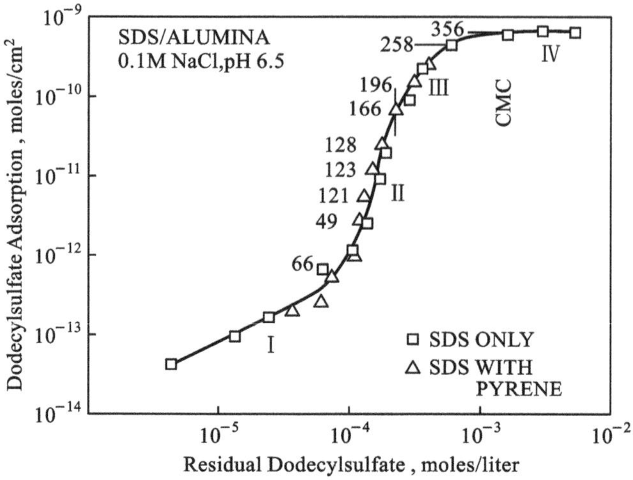

Fig. 4.19. Surfactant aggregation numbers determined at various adsorption densities (average number is indicated).

4.2.1.1.1. *Polymer conformation in the adsorbed state* The nature of adsorption of polymeric molecules on to solid surfaces can be quite different from the adsorption of small molecules (Morawetx, 1975). This stems from the widely varying sizes and configurations available for the adsorptive polymer. In addition, macromolecules usually possess multifunctional groups each having a potential to adsorb on some surfaces.

Among polymeric materials, polyelectrolytes are the most important because of their participation in many biological processes and their utility in industrial processes like dispersion, flocculation, deposition, dewatering, etc.

Polyacrylic acid can exist in different conformations depending on the solvent, pH, and ionic strength conditions (Fig. 4.20) (Aiora and Turro, 1987). Such flexibility also influences its adsorption characteristics on solids and in turn affects subsequent dispersion stability. By using a fluorescent probe labeled polymer and monitoring the extent of excimer formation, it has been shown that the polymer could have a stretched or coiled conformation at the interface depending on the pH. Such studies also demonstrated that the polyacrylic acid adsorbed in a stretched form on alumina at high pH is essentially irreversible as the conformation could not be altered in this case in short order of time by lowering the pH subsequently (Fig. 4.21) (Chandar et al., 1987). In contrast, the polymer adsorbed in the coiled form at low pH stretches out when the pH is increased. In an investigation on the flocculation behavior of alumina particles with polymer under fixed and shifted pH conditions, the polymer conformation was shown to be a controlling factor of the flocculation process. The results of fluorescence emission studies using pyrene labeled PAA in the adsorbed state under shifted pH conditions (i.e. adsorbing the polymer at a fixed pH and then changing the pH to a desired value) is shown in Fig. 4.22. It may be noted that the excimer fluorescence emission intensity of the S/L mixture after adsorption at high pH values remains the same even after changing the pH to lower values. From the obser-

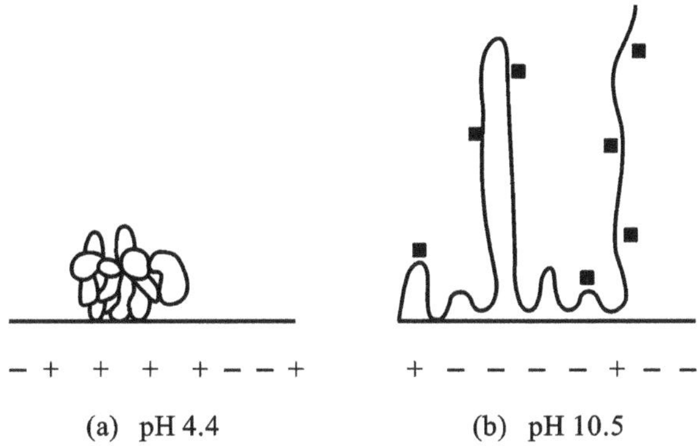

Fig. 4.20. Schematic representation of the conformation of polyacrylic acid at different pH.

vations, the variation of polyacrylic acid conformation at the solid–liquid interface under shifted pH conditions may be represented as shown in Fig. 4.23. It may be inferred that the polymer conformation at a given pH may be manipulated by controlling the adsorption conditions.

4.2.1.2. *Electron spin resonance (ESR)*

ESR studies as applied to micellar systems rely on the sensitivity of a free radical probe to its microenvironment. Molecular species with a free electron posses labeled intrinsic angular momentum (spins), which in an external magnetic field undergoes Zeeman splitting. For a system with $S = 1/2$, two Zeeman energy levels are possible whose energy gap (ΔE) is given by:

$$E = h\nu = gBH_0 \tag{4.19}$$

The magnetic moment of the free electron is also susceptible to the secondary magnetic moments of the nuclei and thus the Zeeman splitting will be superimposed by the hyperfine splitting which brings about further splitting of the absorption signal. The hyperfine splitting pattern depends on the spins and the actual number of the neighboring nuclei with spins. If the electron is in the field of a proton, then the ESR spectrum would yield two lines of equal intensity and similar interaction by a nucleus with $S = 1$, as in nitrogen, would produce a triplet of equal intensity. The line shapes of ESR signals are subject to various relaxation processes occurring within the spin system as well as anisotropic effects due to the differentially oriented paramagnetic centers being acted upon by an external magnetic field (Wertz and Bolton, 1986). These effects result in a broadening of the absorption lines. Three types of ESR studies can be applied to probe the surfactant microstructures—spin probing, spin labeling and spin trapping (Ranby, 1977). In the spin-probing technique, a molecule with a spin is externally added to the system, whereas in spin labeling a spin bearing moiety through covalent bonding forms a part of the molecule.

Information on micropolarity and microviscosity can be obtained by measuring the hyperfine splitting constant A_N and the rotational correlation time τ_c. The hyperfine splitting

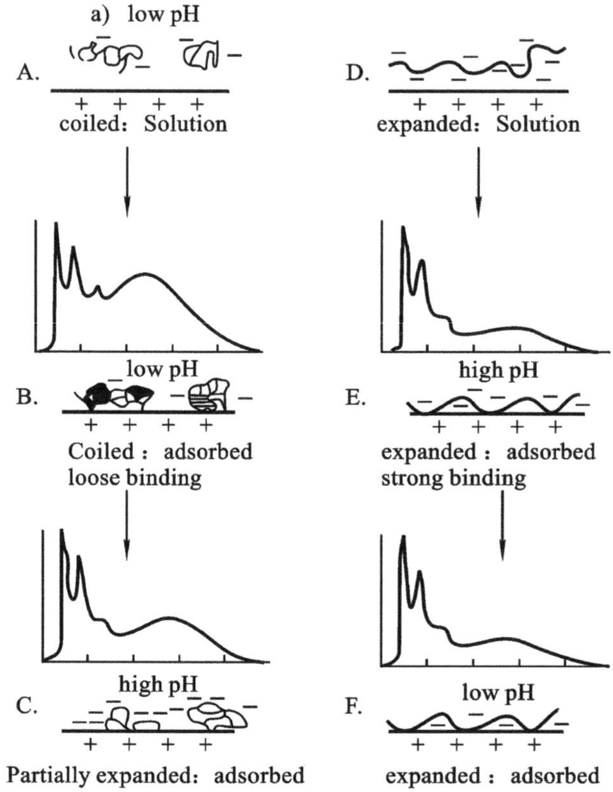

Fig. 4.21. Schematic representation of the adsorption process of pyrene-labeled polyacrylic acid on alumina. (A) At low pH, polymer is coiled in solution which leads to (B) adsorption in coiled form. (C) Subsequent raising of the pH causes some expansion of the polymer. (D) Polymer at high pH in solution is extended and binds (E) strongly to the surface in this conformation. (F) Subsequent lowering of pH does not allow for sufficient intrastrand interactions for coiling to occur.

constants of 16-doxyl stearic acid measured in dodecyl sulfonate solloids (hemi-micelles) (15.0 G) is indicative of a less polar environment compared to its value in water (16.0 G) and SDS micelles (15.6 G) (Fig. 4.24). Similarly microviscosities estimated from τ_c measurements and calibrated against τ_c measured in ethanol–glycerol mixtures give reasonably high values for the solloids compared to those for water. Three different microviscosities were obtained using different probes in the solloids indicating that the nitroxide group in each case felt a different viscosity within the solloid. These observations may be explained by assuming a model for the adsorption of the probe in which the carboxylate group is bound to the alumina surface. Such a model would necessitate attribution of greater mobility near the SDS/H_2O for the nitroxide moiety interface of the 16-doxyl stearic acid case and less mobility of 5-doxyl and 7-doxyl cases (Waterman et al., 1986; Chandar et al., 1986).

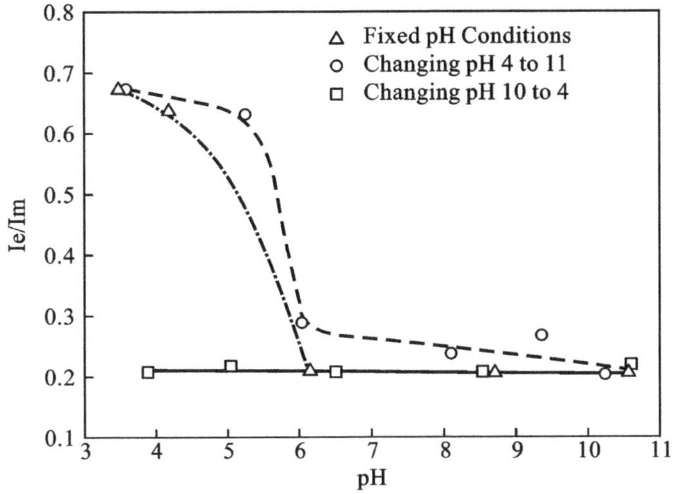

Fig. 4.22. Coiling presented in terms of excimer to monomer ratio, I_e/I_m, for alumina with 20 ppm polyacrylic acid as a function of final pH under shifted pH conditions.

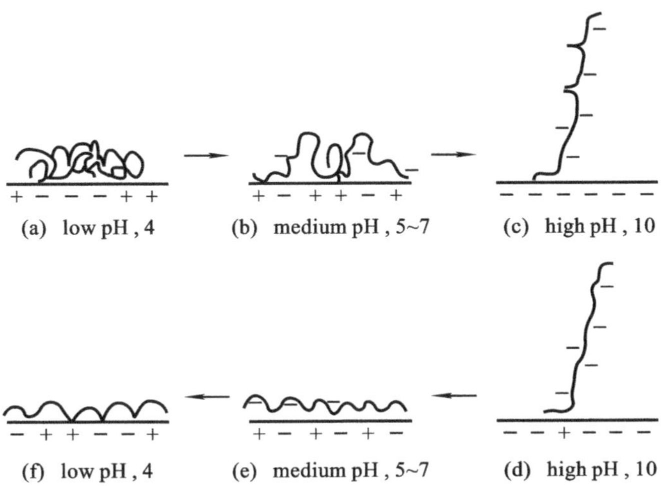

Fig. 4.23. Schematic representation of the variation of polymer conformation at the solid–liquid interface under shifted pH conditions.

4.2.1.3. *Raman spectroscopy*

The alumina/dodecylsulfonate system has also been probed by excited state resonance Raman spectroscopy using tris(2,2′-bipyridyl) ruthenium(II) chloride, $Ru(bpy)_3^{2+}$ as a probe molecule (Somasundaran et al., 1989). It has been shown that ruthenium polypyridyl complexes serve as excellent photophysical probes for biopolymers like nucleic acids. The excited state of $Ru(bpy)_3^{2+}$ shows strong resonance enhanced Raman transitions when probed at 355 nm^{-1}. Furthermore, it has been shown that binding of these ions to clay

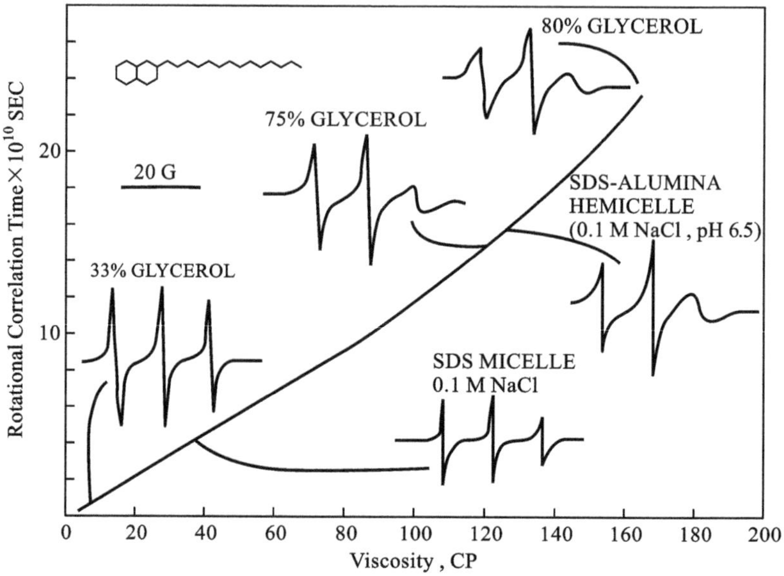

Fig. 4.24. ESR spectra of 16-doxyl stearic acid in solloids (hemimicelles), micelles and ethanol–glycerol mixtures and comparison of corresponding rotational correlation times and viscosities.

Table 4.1
Normalized intensities for various Raman transitions in aqueous solutions and SDS micelles

Peak position (cm^{-1})	Normalized intensities	
	Aqueous solution (±3%)	SDS micelles (±3%)
1213	0.66	0.84
1286	1.00	1.00
1324	0.11	0.20
1426	0.42	0.42
1499	0.47	0.73
1547	0.85	0.92
1605	0.26	0.39

particles results in substantial changes in the ground-state transitions of the spectrum. For these reasons this probe was chosen to study the solloids formed at the alumina–water interface with excited-state resonance Raman spectroscopy.

Raman spectra of $Ru(bpy)_3^{2+}$ above the CMC show frequency shifts as well as intensity changes as compared to its spectrum in water. Table 4.1 lists the relative intensities of various lines with respect to the line at 1286 in water. These transitions can be attributed to the perturbation of the excited state by the SDS micelles. Excited state resonance Raman spectrum of this probe in various regions of the adsorption isotherm for the alumina/SDS system is shown in Fig. 4.25. The spectrum of $Ru(bpy)_3^{2+}$ on alu-

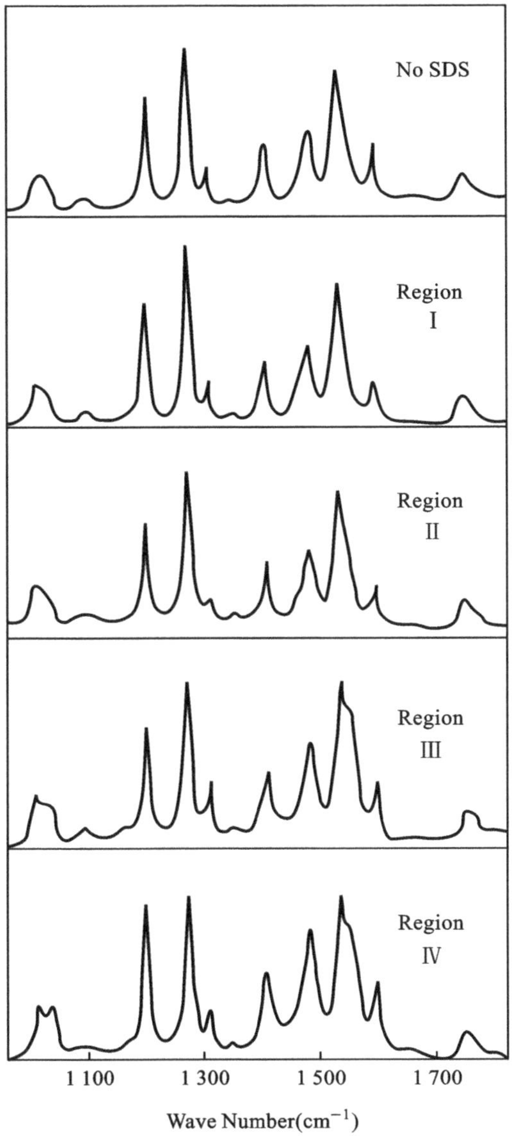

Fig. 4.25. Resonance Raman spectrum of Ruby(bpy)$_3^{2+}$ for various regions of the alumina/SDS adsorption system.

mina in the absence of SDS is very much similar to its spectrum in water both in terms of frequencies and relative intensities. This trend is continued into Region II, where the solloid aggregation process starts. In Regions III and IV, the Raman spectrum shows dramatic changes and the frequency shifts are more pronounced than in the case of micelles. A plot of the change in wave numbers for some of the lines in the

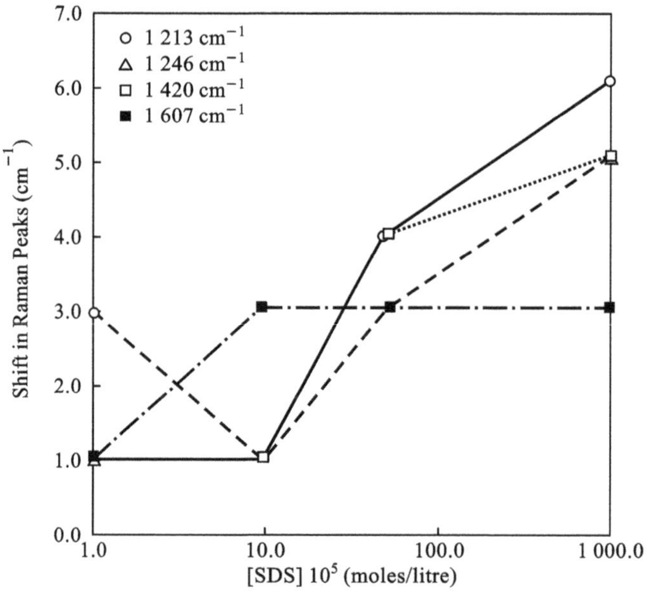

Fig. 4.26. Frequency shifts of resonance Raman lines of $Ru(bpy)_3^{2+}$ as a function of SDS concentration for an alumina/SDS system.

four different regions of the adsorption isotherm is shown in Fig. 4.26. The changes in Raman frequency and intensity assume substantial significance in the transition Regions II and III onwards only. This could be due to the change of net charge on the alumina surface from positive to negative. The favorable net negative charge enhances the adsorption of the probe at the solid–liquid interface. Accordingly, no adsorption of $Ru(bpy)_3^{2+}$ was observed when the supernatants were analyzed in Region I and Region II or in the absence of SDS. The transitions at 1213, 1286 and 1428 cm^{-1} show significant increments and these trends clearly suggest probe adsorption on to the solloids. Also it was seen that the 1286 peak shifted to 1281 in the presence of solloids while it was practically unchanged in SDS micelles. It may be speculated that $Ru(bpy)_3^{2+}$ may be sensing different environments within the SDS micelles and the alumina/SDS solloids.

4.3. SURFACE AND BULK INTERACTIONS BETWEEN DISSOLVED MINERAL SPECIES AND FLOTATION AGENTS

Minerals such as calcite, apatite and wolframite that are sparingly soluble in aqueous solutions will release dissolved species. In addition, these dissolved mineral species can also interact with various minerals, resulting in chemical alterations of their surfaces which can in turn interact with various flotation agents added. All such interactions can lead to a loss of selectivity in the flotation operation. These phenomena based upon solution

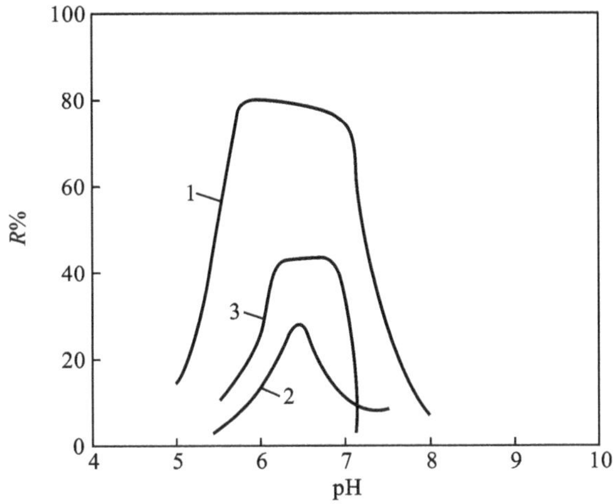

Fig. 4.27. Floatability of apatite with oleate (2×10^{-6} mol/l, $I = 3 \times 10^{-2}$ mol/l KNO^3) in: 1—pure water; 2—saturated calcite supernatant; 3—saturated apatite supernatant.

equilibria of all the minerals and the flotation agents in the system are much more complex than those of the individual systems. It is important to have a full understanding of these since they determine essentially the efficiency or selectivity of flotation operations.

Various types of flotation agents are used for selective hydrophobization of minerals in froth flotation. Oleic acid is the most common collector used commercially in the flotation processes of salt-type mineral because of its low cost and availability. In the case of complex ores containing similar constituent minerals such as calcium carbonate and calcium phosphate, the selectivity of flotation using oleate as a collector is usually poor and the reagent consumption is generally high (Lawver et al., 1982; Hanna and Somasundaran, 1976). In order to achieve improvements in the separation, it is helpful to identify various interactions between the dissolved mineral species and the oleate species. These interactions are discussed below.

4.3.1. Competitive adsorption

Competition of the dissolved mineral species with similarly charged collector species for adsorption sites can result in the depression of mineral flotation. For example, in the case of flotation of apatite and calcite with oleate as collector, the floatability is usually lower in saturated mineral supernatants than in pure water (Fig. 4.27). This is partly due to the competition by dissolved species such as $P_3O_4^{2-}$ and CO_3^{2-} which are present in saturated apatite and calcite supernatants (10^{-6} to 10^{-4} mol/l) in concentrations close to that of the anionic oleate ions. Similar effects are also observed in the case of flotation of fluorite with dodecyl amine (Fig. 4.28). Floatability of fluorite in its saturated solution is lower than that in water. The concentration of dissolved Ca species in the fluorite saturated solution is about 10^{-4} mol/l and only when the amine concentration is higher than 10^{-4} mol/l, the flotation obtained in it is similar to that obtained in water.

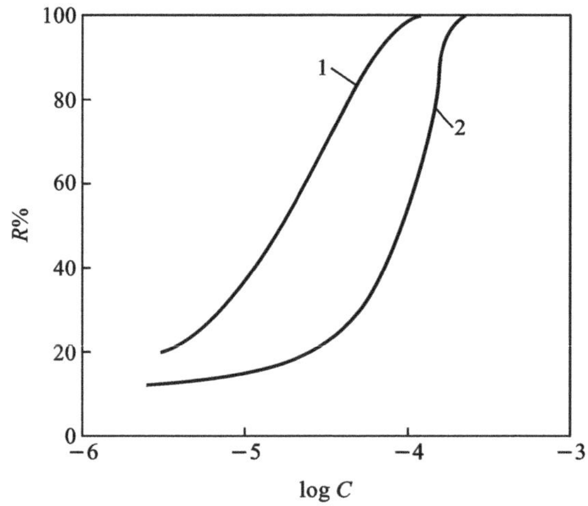

Fig. 4.28. Floatability of fluorite with dodecyl amine in: 1—pure water; 2—saturated fluorite supernatant.

4.3.2. Surface and bulk precipitation

Solution equilibria of systems containing both minerals and flotation agents are rather complex. In addition to the dissolution of minerals and adsorption of dissolved mineral and reagent species at the interface, interactions between these species in the bulk can lead to complicating phenomena such as surface and bulk precipitation as well as redissolution of adsorbed species. It is to be noted that the physico-chemical nature of adsorption processes is in many ways similar to that of bulk reactions. In a number of cases, the solution conditions for the so-called chemisorption correspond to that for the precipitation of the metal-reagent salts in the interfacial region, suggesting that chemisorption may in fact be a surface precipitation phenomenon. This is illustrated in the following section with dolomite–francolite–oleate system as an example. A multi-pronged approach involving adsorption, microflotation, dissolution and spectroscopy tests was used to identify various surface and bulk interactions (Xiao, 1990).

4.3.3. Floatability, adsorption and dissolved species

Floatabilities of francolite and dolomite using oleate as collector are shown in Figs. 4.28 and 4.29 as a function of oleate concentration and pH. It can be seen that the floatabilities of dolomite and francolite are quite similar with a sudden increase at an oleate concentration of 1.6×10^{-4} mol/l. Selective flotation of francolite from dolomite might be expected only above pH 9 or below pH 5. However, flotation of a binary mixture of dolomite and francolite (50:50) with 1.7×10^{-4} mol/l oleate at pH 5.2 and 9.0 yielded poor separation. The loss of selectivity during flotation may be attributed to the interactions among the dissolved mineral species from minerals and oleate as well as the formation of oleate precipitates (Amankonah and Somasundaran, 1985). Depletion isotherms for ^{14}C labeled oleate on francolite and dolomite are shown in Fig. 4.30. Both francolite and dolomite show a two-region (II and III) linear isotherms with a break at a residual concentration of

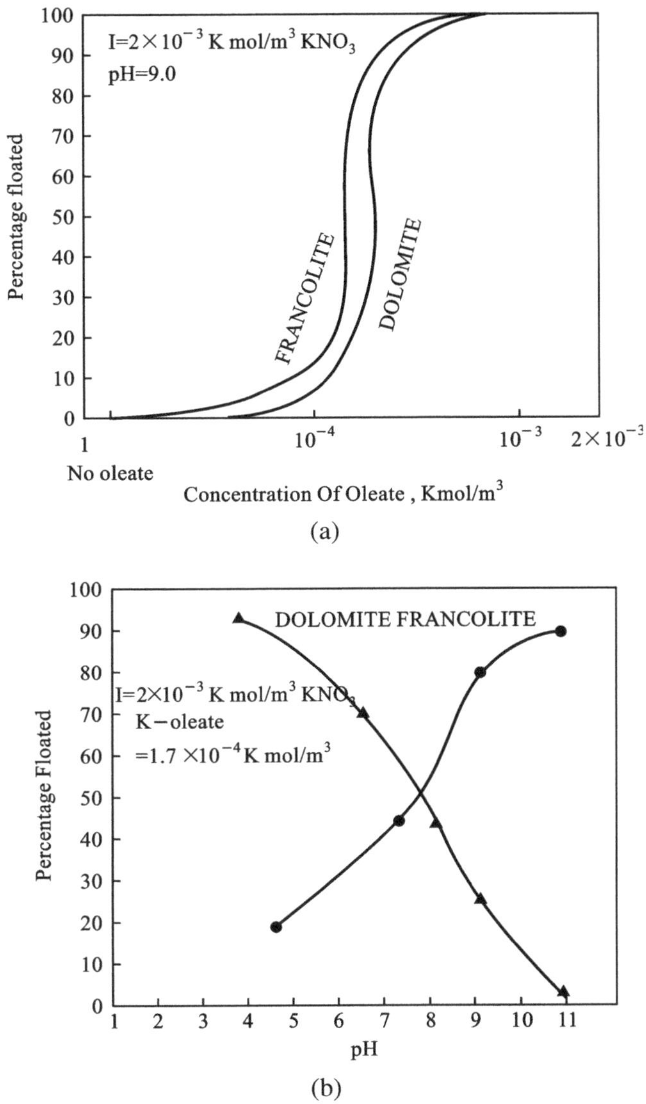

Fig. 4.29. (a) Flotation of francolite and dolomite as a function of oleate concentration at pH = 9.0; (b) flotation of francolite and dolomite as a function of pH at oleate concentration of 1.7×10^{-4} mol/l.

10^{-5} mol/l. At a concentration of $<10^{-5}$ mol/l (Region II), both curves show a linear increase resulting from the oleate adsorption on the particles due to electrostatic interaction. The following linear equation was obtained for both of the isotherms:

$$\Gamma = 0.824C - b, \tag{4.20}$$

where Γ is adsorption density of oleate; C is residual concentration; and b is the intercept of the isotherms and equals to 1.65×10^{-6} for dolomite and 2.88×10^{-6} for francolite.

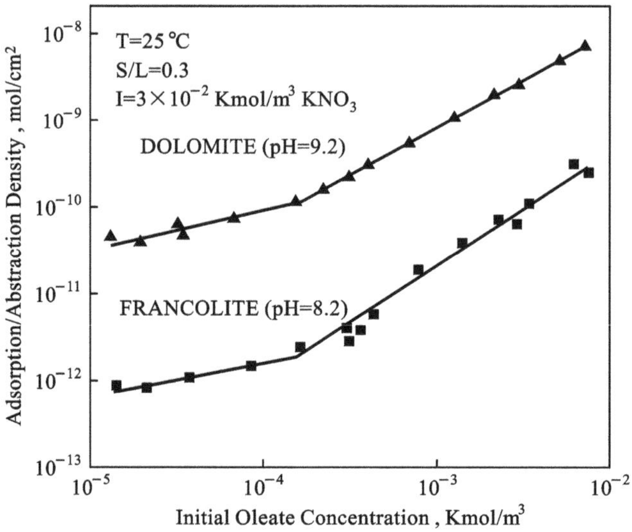

Fig. 4.30. Depletion isotherms of 14C labeled oleic acid on francolite and dolomite.

At a concentration of $>10^{-5}$ mol/l (Region II), the slope of both curves increases significantly. This increased adsorption is attributed to possible precipitation of oleate on the mineral surface as well as in the bulk solution. The equation for both the isotherms is given below:

$$\Gamma = 0.923C - b, \tag{4.21}$$

where $b = 1.38 \times 10^{-5}$ for dolomite and 8.24×10^{-5} for francolite.

Molecular interactions in mineral/oleate systems can be investigated by monitoring dissolved mineral species in the supernatants of the samples used in the adsorption experiments. The data obtained for dissolved Ca and Mg species is shown in Fig. 4.31 and both plots can be divided into three regions. At low concentrations of oleate (below 10^{-5} mol/l (Region I)), the lines are almost horizontal indicating the absence of precipitation of calcium and magnesium. Above 10^{-5} mol/l in Region II, Ca and Mg levels in the supernatant gradually decrease for both francolite and dolomite and above 10^{-4} mol/l in Region III, Ca and Mg levels were found to decrease sharply. This suggests that bulk precipitation of calcium and magnesium species with oleate leading to severe depletion of Ca and Mg species occurs in the concentration range in which abrupt transition from Region II to III was observed on the oleate adsorption isotherms (Fig. 4.30).

Relevant chemical equilibria responsible for the precipitation of Ca and Mg species with oleate are as follows (du Rietz, 1975; Leja, 1982):

$$Ca^{2+} + 2Ol^{-} = Ca(Ol)_2 \qquad K_{spCa(Ol)_2} = 3.81 \times 10^{-13}$$

$$CaOH^{+} + 2Ol^{-} = Ca(Ol)_2 + OH^{-} \tag{4.22}$$

$$Mg^{2+} + 2Ol^{-} = Mg(Ol)_2 \qquad K_{spMg(Ol)_2} = 1.58 \times 10^{-11}$$

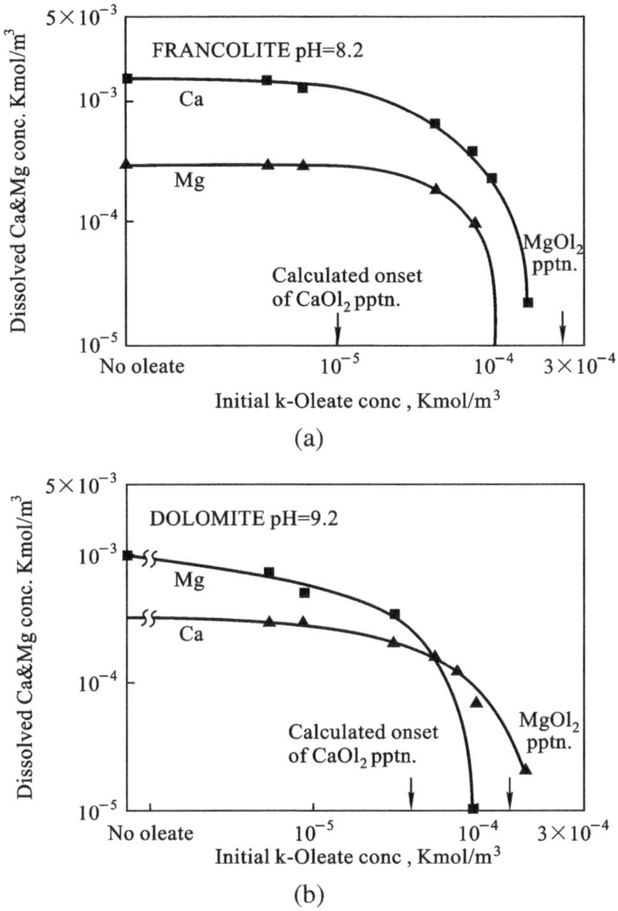

Fig. 4.31. Dissolved Ca and Mg levels from supernatants of: (a) francolite and (b) dolomite as a function of oleate concentration.

$$MgOH^+ + 2Ol^- = Mg(Ol)_2 + OH^- \qquad (4.23)$$

Formation of calcium and magnesium oleates on the mineral surface as well as in the bulk solution results in surface and bulk precipitations. The concentrations of onset of the precipitation of $CaOl_2$ and $MgOl_2$ are calculated from their solubility products for dolomite–oleate and francolite–oleate systems, and these are marked in Figs. 4.31a and 4.31b. The oleate concentrations at the onset of precipitation calculated from the solubility data for $Ca(Ol_2)$ and $Mg(Ol_2)$ are in good agreement with those obtained from dissolved species analysis (Fig. 4.31). It can also be seen that the onset of surface precipitation (transition between Regions II and III in Fig. 4.31) corresponds to the concentration region in which flotation of both the minerals shows a sharp increase (Fig. 4.29), indicating that the hydrophobicity of these minerals is induced predominantly by the surface precipitation of calcium and magnesium oleates. A further increase in the oleate concentration results in

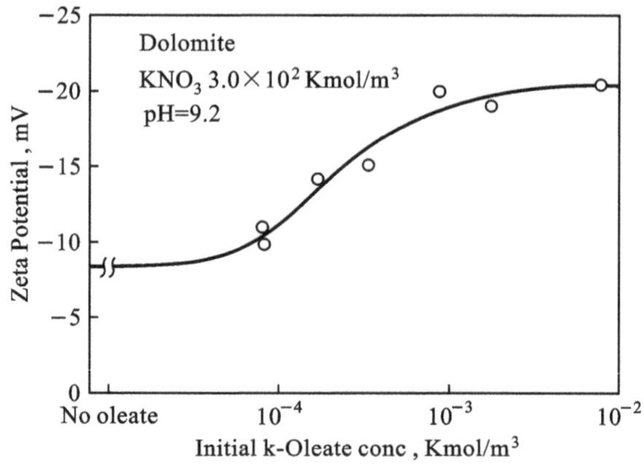

Fig. 4.32. Zeta-potential of dolomite in the presence of oleate.

the precipitation of calcium and magnesium oleates in the bulk solution. Such bulk precipitates may coat the surface of both the minerals making their interfacial properties similar and this will indeed lead to a loss of selectivity in flotation.

4.3.4. Characterization of surface and bulk precipitates

As discussed earlier, addition of oleate to a suspension of francolite or dolomite can result in both surface and bulk precipitation of calcium and magnesium oleate complexes. Also, the flotation onsets of dolomite and francolite (around 1.0×10^{-4} mol/l in Fig. 4.29) are in the region in which surface precipitation of Ca and Mg oleates takes place (Fig. 4.31). This suggests that the hydrophobicity required for good flotation is obtained in this case by the surface precipitation of calcium and/or magnesium oleates. However, at higher oleate concentrations calcium and magnesium oleate bulk precipitation occurs, which can coat particles of both the minerals and cause them to respond similarly in flotation resulting in poor selectivity. Presence of surface and bulk precipitates in the system has been confirmed using in-situ experimental techniques such as zeta-potential, FTIR spectroscopy and nephelometry. Fig. 4.32 shows the zeta-potential of dolomite as a function of oleate concentration. The zeta-potential exhibits an increase (of the negative value) in the range 10^{-4} to 10^{-3} mol/l of oleate concentration. This suggests that the adsorption of anionic oleate species on the mineral surface makes the surface more negative, but further increase in oleate depletion resulting from bulk precipitation does not change the zeta-potential of the mineral. This is further examined by ATR/FTIR spectroscopy. The ATR spectra of dolomite slurry in various oleate solutions corresponding to the three regions discussed earlier (Fig. 4.29) are shown in Fig. 4.33. It can be seen that the two peaks at 1448 and 1351 cm^{-1}, corresponding to the stretching vibration of C=O bond in the carbonate groups and the C–O stretching vibration of the C–O groups in Ca-oleate precipitates formed, respectively, change with the oleate concentration. In the presence of the oleate, the intensity of the carbonate peak at 1448 cm^{-1} is diminished with the increase in the oleate concentration. This is due to the formation of Ca and Mg oleate precipitates and their exchange

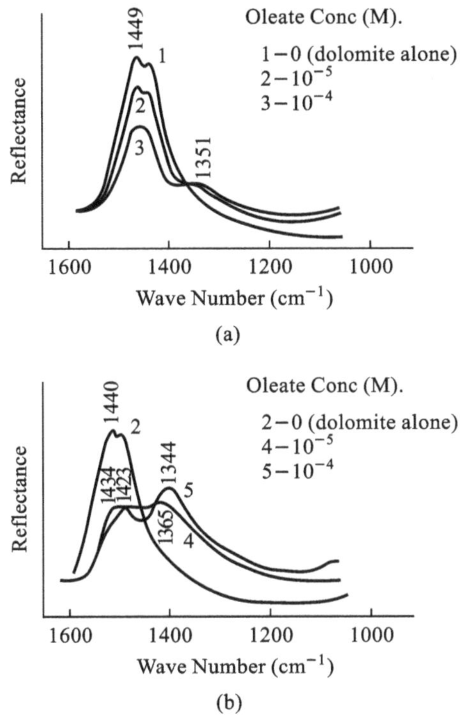

Fig. 4.33. ATR spectra of dolomite slurry at different concentrations of oleate ((a) and (b)) solutions.

with the carbonate groups on the mineral surface. Possible chemical equilibria involved in this are as follows:

$$CaMg(CO_3)_{2(s)} + 4RCOO^- = Ca(RCOO)_{2(s)} + Mg(RCOO)_{2(s)} + 2(CO_3)^{2-} \quad (4.24)$$

or

$$CaCO_{3(s)} + 2RCOO^- = Ca(RCOO)_{2(s)} + (CO_3)^{2-}$$

$$MgCO_{3(s)} + 2RCOO^- = Mg(RCOO)_{2(s)} + (CO_3)^{2-} \quad (4.25)$$

According to the above reactions, removal of carbonate groups from the mineral surface should result in the reduction of the intensity of carbonate peak as shown in Fig. 4.33. On the other hand, adsorption and surface precipitation of oleate can take place on the dolomite surface leading also to similar reductions. A new peak at 1351 cm^{-1} when the oleate concentration reaches about 10^{-4} mol/l can be clearly seen in Fig. 4.33b and the intensity of this peak increases with the increase in the oleate concentration, where precipitation of calcium and magnesium oleates is likely to occur both on the particle surface and in the bulk. The peak heights of both the carbonate and the oleate (I_{co} and I_{ol}) in the spectra were measured. Bulk precipitation using a common tangent baseline method (Willis et al., 1987) and the ratio I_{co}/I_{ol}, is plotted in Fig. 4.34. This plot also

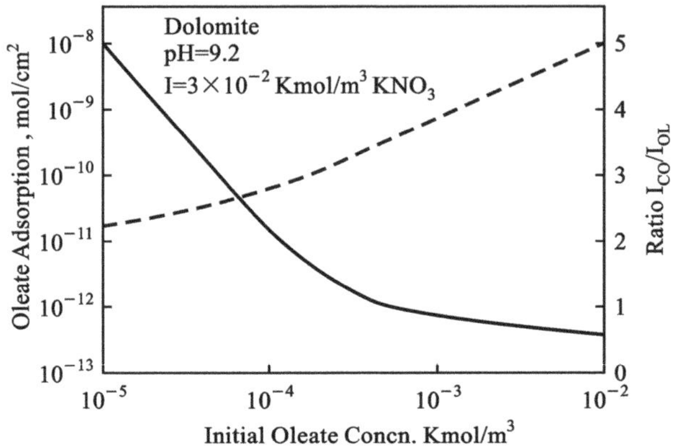

Fig. 4.34. Adsorption and I_{co}/I_{ol} ratio from ATR spectra of dolomite–oleate system.

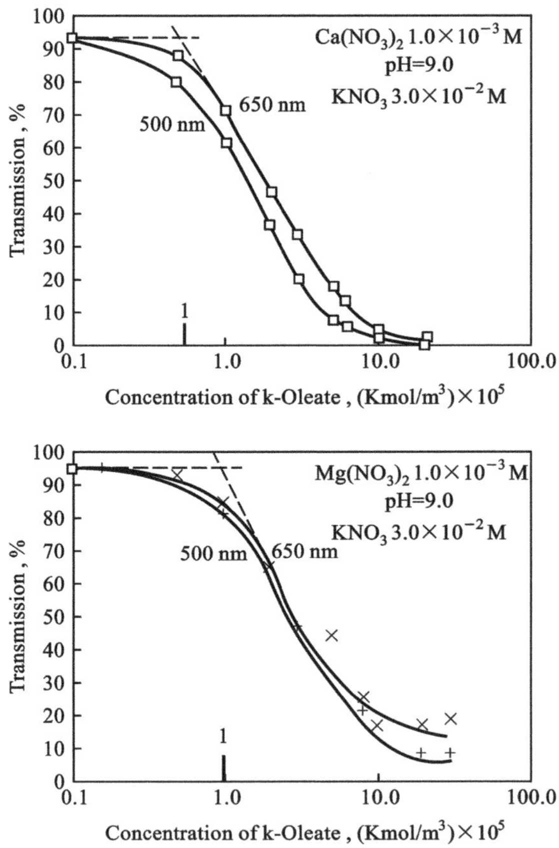

Fig. 4.35. Transmission through: (a) Ca nitrate and (b) Mg nitrate solutions at pH 9 as a function of oleate concentration at wavelengths of 500 and 650 nm.

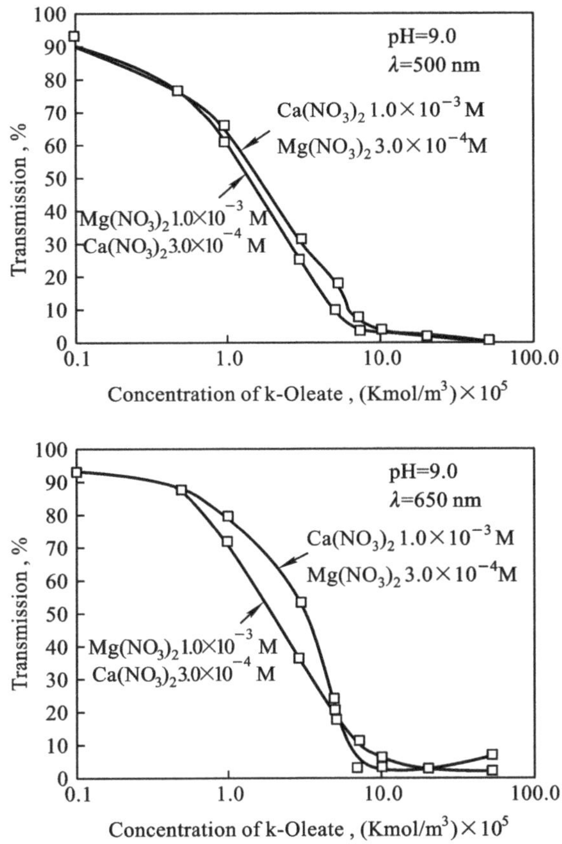

Fig. 4.36. Transmission through Ca and Mg nitrate solutions at pH 9 as a function of oleate concentration at wavelengths of: (a) 650 nm and (b) 500 nm.

shows a marked change in the slope of I_{co}/I_{ol} curve at about 2×10^{-4} kmol/m^3 of oleate, indicating a significant change in the surface properties of the mineral that might be attributed to oleate precipitation of calcium and magnesium oleates monitored simultaneously by determining the concentrations of dissolved Ca and Mg species by atomic absorption and the transmittance of calcium and magnesium nitrate solutions with added oleate. Transmittance through the colloidal suspensions of calcium oleate and magnesium oleate is shown in Fig. 4.35 as a function of the initial concentration of oleate. The onset of precipitation of calcium oleate (5.2×10^{-6} mol/l) obtained from the graph is lower than that of magnesium oleate (1.0×10^{-5} mol/l oleate) and this was attributed to the difference between the solubility products of calcium oleate (3.8×10^{-13}) and magnesium oleate (1.6×10^{-11}). Since francolite and dolomite contain both dissolved calcium and magnesium species in the supernatant, the effect of one component in the presence of the other on the oleate precipitation is also examined. Fig. 4.36 shows the transmission curves of calcium and magnesium oleate solutions for the addition of Mg

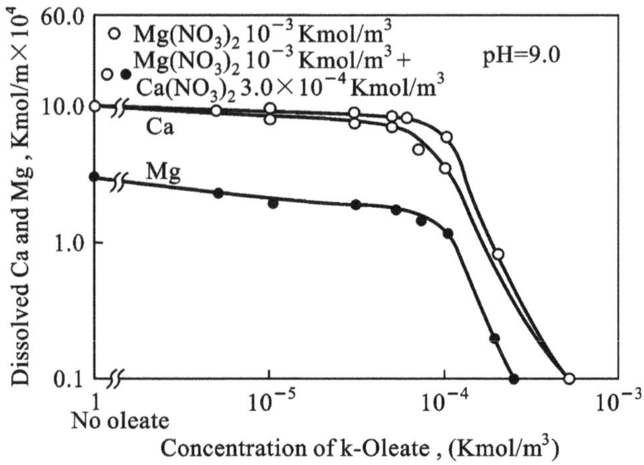

(a)

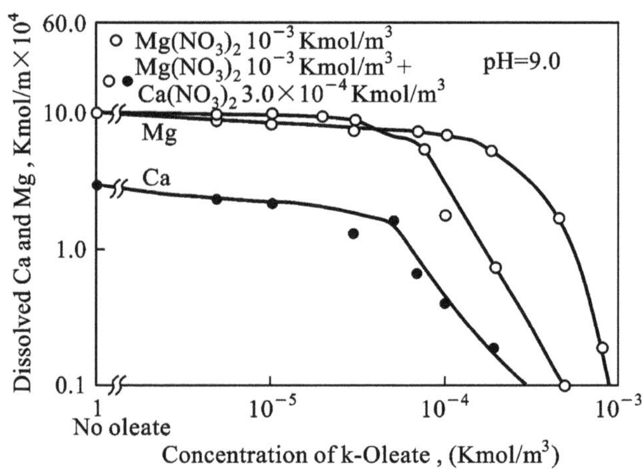

(b)

Fig. 4.37. Dissolved Ca and Mg levels of: (a) Ca and Ca/Mg nitrate mixtures and (b) Mg and Ca/Mg nitrate mixtures as a function of oleate concentration at pH 9.

species as magnesium nitrate (3.0×10^{-4} mol/l to simulate francolite supernatant), and that of Ca species as calcium nitrate (3.0×10^{-4} mol/l to simulate dolomite supernatant).

It can be seen that magnesium addition to calcium oleate solution or vice versa does shift the transmission curve to lower oleate concentrations, indicating co-precipitation of calcium and magnesium oleates due to synergistic effects. Overlapping of both the curves in Fig. 4.36 is also in line with the behavior of the dolomite and francolite systems in microflotation and dissolution experiments. Simultaneous measurements of dissolved species in the supernatant of the precipitated solutions (Fig. 4.37) showed Ca and Mg levels to remain almost constant until solubility products of their oleate are reached (2×10^{-4} mol/l for calcium nitrate and 4×10^{-4} mol/l for magnesium nitrate). When a mixture of calcium nitrate and magnesium nitrate is present, the onset of precipitation of either calcium or magnesium with oleate is shifted to a lower concentration of oleate, again indicating synergism for co-precipitation of Ca and Mg species with oleate. In the presence of a mixture of 10^{-3} mol/l $Ca(NO_3)_2$ and 3.0×10^{-4} mol/l $Mg(NO_3)_2$ (simulating francolite supernatant), the precipitation of calcium oleate is not affected significantly by the addition of magnesium but the precipitation of Mg-oleate is affected by calcium. Similarly, in the presence of a mixture of 10^{-3} M $Mg(NO_3)_2$ and 3.0×10^{-4} M $Ca(NO_3)_2$ (simulated dolomite supernatant), the onset of precipitation of magnesium oleate was shifted to lower concentration of oleate by the addition of calcium, but calcium oleate precipitation itself was not affected. Above results in pure solutions can be compared with flotation results for dolomite/francolite flotation systems (Fig. 4.1). The shifts in the onsets of precipitation and changes in the shape of the curve can be attributed to adsorption and surface precipitation of oleate species on the mineral surface.

4.3.5. Molecular model of oleate on francolite and dolomite

Overall behavior of the dolomite/francolite–oleate flotation system is depicted schematically in Fig. 4.38 and its implications are discussed below:

1. At low concentrations ($<10^{-5}$ mol/l, Region I), oleate species adsorb individually on the mineral possibly due to electrostatic interactions. In this range, no precipitation of dissolved calcium and magnesium species with oleate occurs.
2. At intermediate concentrations (10^{-5} to 2×10^{-4} mol/l, Region II), the solubility limit of calcium and magnesium oleates may exceed in the interfacial region but not in the bulk solution. Then, surface precipitation of oleate on the mineral particles occurs leading to an increase in oleate depletion and a decrease in the levels of dissolved calcium and magnesium species. The hydrocarbon tails of the oleate molecules would be oriented towards the bulk making the mineral surface hydrophobic.

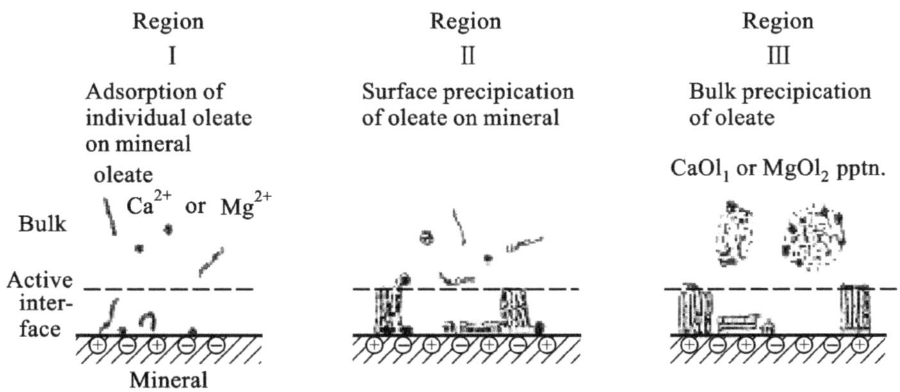

Fig. 4.38. Molecular model of interaction of oleate species at mineral–solution interface.

3. At higher concentrations (>2.0 × 10^{-4} mol/l, Region III), the slope of the isotherm is higher and a sharp decrease in concentrations of dissolved species takes place. This suggests that oleate depletion and bulk precipitation of calcium and magnesium oleates predominates in this region.

4.4. SOLUTION CHEMICAL REACTIONS AND DIAGRAMMATIC APPROACHES

In 1940's, Taggart (Taggart et al., 1930) proposed that flotation behavior of minerals is determined by chemical reactions between reagents and metallic species of minerals in solution. The solubility products of reaction compounds is normally used as the criteria.

4.4.1. Competitive reaction between collector ion and hydroxyl ions in sulfide flotation

In the flotation pulp of sulfide minerals, the reaction between collector and metallic salt can be given as follows:

$$M^{2+} + 2X^- = MX_2, \qquad K_{spMX_2} = [M^{2+}]\,[X^-]^2 \tag{4.26}$$

The reaction forming hydroxides is:

$$M^{2+} + 2OH^- = M(OH)_2, \qquad K_{spM(OH)_2} = [M^{2+}]\,[OH^-]^2 \tag{4.27}$$

Thus, the total reaction is:

$$MX_2 + 2OH^- = M(OH)_2 + 2X^- \tag{4.28}$$

The equilibrium constant for the above reaction is

$$K' = [X^-]^2/[OH^-]^2 = K_{spMX_2}/K_{spM(OH)_2} \tag{4.29}$$

or

$$K = (K')^{1/2} = [X^-]/[OH^-] = (K_{spMX_2}/K_{spM(OH)_2})^{1/2} \tag{4.30}$$

K is called the Barsky's constant, which shows competitive reaction between reagent and hydroxyl ions in flotation. For example, the flotation of pyrite, galena and chrysocolla by ethyl xanthate as a function of pH is shown in Fig. 4.39.

According to Eq. (4.30), the relation for H and collector ions is:

$$[X^-][H^+] = (K')^{1/2} \times 10^{-14} = (K_{spMX_2}/K_{spM(OH)_2})^{1/2} \times 10^{-14} \tag{4.31}$$

For galena–ethyl xanthate system, the reaction constants are:

$$K_{spPbX_2} = 10^{-16.7}, \qquad K_{spPb(OH)_2} = 10^{-15.1} \tag{4.32}$$

In addition, the dissociation constant of ethyl xanthatic acid is $K_a = 10^{-2}$–10^{-5}. The concentration of xanthate is equal to the total concentration of xanthate in an alkaline medium.

$$[X^-][H^+] = (10^{-16.7}/10^{-15.1})^{1/2} \times 10^{-14} = 1.58 \times 10^{-15} \tag{4.33}$$

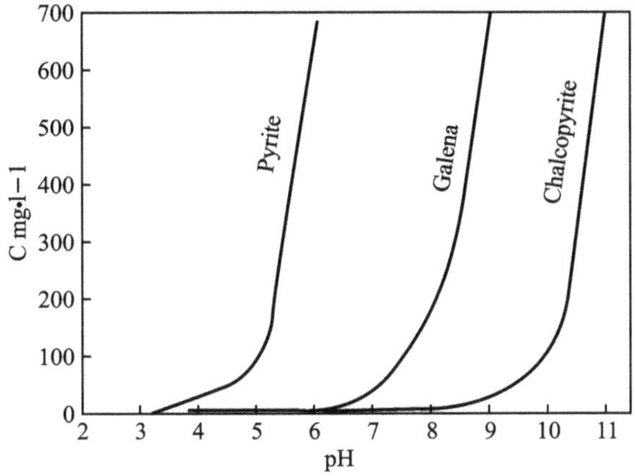

Fig. 4.39. Concentration of ethyl xanthate required for the flotation of pyrite, galena and chalcoprite (Wark and Cox, 1934).

As an example, the data obtained from the Sullivan concentrator of Canada for galena flotation is given below (Bushell, 1958):

$[X^-][H^+] = 2.0 \times 10^{-16}$–$7.9 \times 10^{-16}$ for rougher flotation, and

$[X^-][H^+] = 1.1 \times 10^{-16}$–$5.0 \times 10^{-16}$ for scavenger flotation

4.4.2. Activation of metallic salt in the flotation of pyrite

The activation of sphalerite flotation by lead has been studied by Fuerstenau et al. (1960):

$$ZnS + Pb^{2+} = PbS + Zn^{2+} \quad (4.34)$$

and its equilibrium constant is:

$$K = K_{spZnS}/K_{spPbS} = [Zn^{2+}]/[Pb^{2+}] = 3.0 \times 10^{-26}/8.0 \times 10^{-28} = 100 \quad (4.35)$$

This is in agreement with the experimental value for ratio, 100, obtained for lead ion to zinc ions adsorbed on sphalerite.

4.4.3. p*C*–pH diagrams

This method is based on the reactions between a reagent and relevant mineral species and using solubility product values. In p*C*–pH diagrams, the abscissa is pH of the solution; the ordinate is negative logarithm of the precipitation concentration of the metallic ions.

4.4.3.1. *Determination of flotation conditions for polymetallic sulfides*

For butyl xanthate–sulfide mineral system, the relevant reactions are as follows:

$$Cu^{2+} + 2BX^- = Cu(BX)_2, \qquad K_{Cu(BX)_2} = 10^{-26.2} \quad (4.36)$$

$$Pb^{2+} + 2BX^- = Pb(BX)_2, \qquad K_{Pb(BX)_2} = 10^{-18.0} \tag{4.37}$$

$$Zn^{2+} + 2BX^- = Zn(BX)_2, \qquad K_{Zn(BX)_2} = 10^{-10.43} \tag{4.38}$$

$$Fe^{2+} + 2BX^- = Fe(BX)_2, \qquad K_{Fe(BX)_2} = 2.7 \times 10^{-11} \tag{4.39}$$

Assuming that the concentration of xanthate ion $[BX^-] = 10^{-3}$ mol/l, and those of metallic species are:

$$pCu^{2+} = 20.0, \qquad pPb^{2+} = 12.0, \qquad pZn^{2+} = 4.43, \qquad pFe^{2+} = 4.0$$

The precipitation reactions of metallic hydroxyl compounds are given as follows:

$$Cu^{2+} + 2OH^- = Cu(OH)_2, \qquad K_{spCu(OH)_2} = 10^{-18.32} \tag{4.40}$$

$$Pb^{2+} + 2OH^- = Pb(OH)_2, \qquad K_{spPb(OH)_2} = 10^{-15.1} \tag{4.41}$$

$$Zn^{2+} + 2OH^- = Zn(OH)_2, \qquad K_{spZn(OH)_2} = 10^{-16.2} \tag{4.42}$$

$$Fe^{2+} + 2OH^- = Fe(OH)_2, \qquad K_{spFe(OH)_2} = 10^{-15.2} \tag{4.43}$$

Thus,

$$\begin{aligned} pCu^{2+} &= -9.68 + 2pH \\ pPb^{2+} &= -12.9 + 2pH \\ pZn^{2+} &= -11.8 + 2pH \\ pFe^{2+} &= -12.9 + 2pH \end{aligned} \tag{4.44}$$

Using Eq. (4.42), the pMe^{2+}–pH diagrams given in Fig. 4.40 can be generated. In this diagram, the horizontal line at the bottom represents the minimum concentration of metallic ions which is necessary for the formation of ethyl xanthate salt; the intersection point of horizontal and vertical lines gives the upper pH limit for the mineral flotation.

4.4.3.2. *Determination of conditions for the flotation of Fe-Ca containing minerals using oleate*

pFe^{2+}–pH and pCa^{2+}–pH diagrams generated by du Rietz for flotation of Fe-Ca containing minerals, are shown in Fig. 4.41. It can be seen that at an oleate concentration of 10^{-5} mol/l and pH > 8, Ca mineral can be floated and Fe mineral depressed, permitting separation of Ca from Fe minerals.

4.4.4. ΔG–pH diagrams

Free energy change, ΔG, for solution reactions between reagents and metallic ions can be used as a criterion to generate ΔG–pH diagrams. Flotation mechanisms and flotation conditions can be estimated using these diagrams.

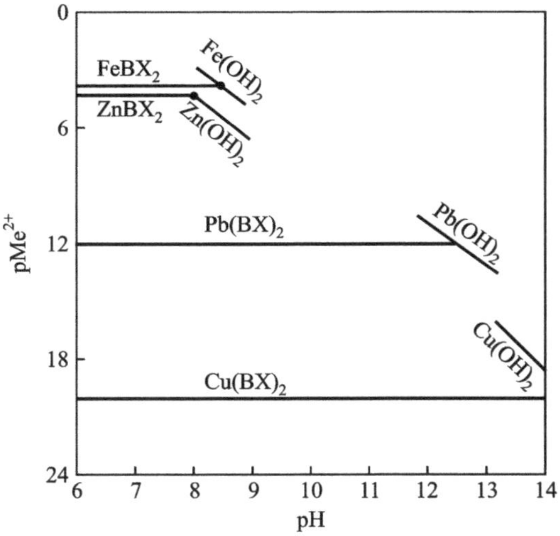

Fig. 4.40. pC–pH diagram of butyl xanthate with Cu^{2+}, Pb^{2+}, Zn^{2+} and Fe^{2+}.

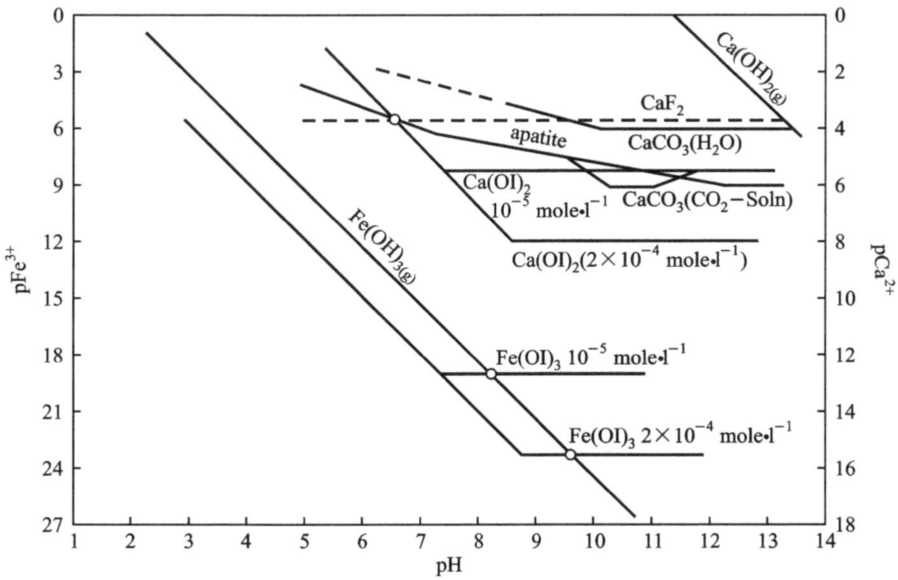

Fig. 4.41. pC–pH diagram of oleate–Fe-Ca mineral system (du Rietz, 1975).

4.4.4.1. *Xanthate–sulfide system*

The reactions between xanthate and sulfide minerals can be given as follows:

$$M^{2+} + 2X^- = MX^2 \tag{4.45}$$

$$\Delta G = \Delta G^0 - RT\ \ln[M^{2+}][X^-]^2 \tag{4.46}$$

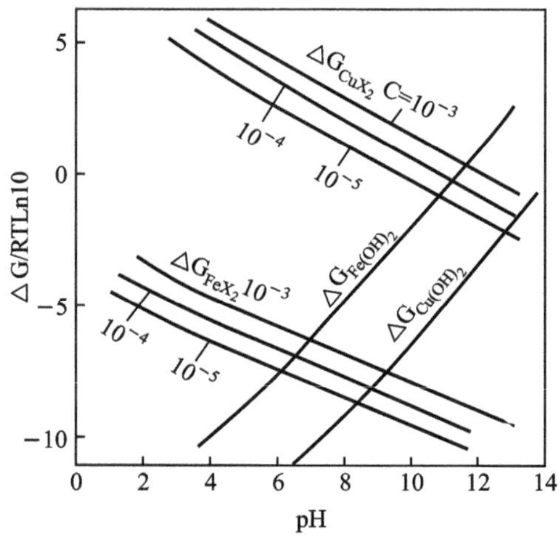

Fig. 4.42. ΔG–pH diagram of Cu and Fe sulfides in the presence of xanthate (Mukai et al., 1962, 1972).

$$[M^{2+}] = 1/\alpha_M (K_{SMS}\alpha_M\alpha_S)^{1/2} = (K_{SMS}\alpha_M/\alpha_S)^{1/2} \tag{4.47}$$

$$[X^-] = [X]/\alpha_X \tag{4.48}$$

$$\Delta G^0 = \mathrm{RT}\ln K_{SMX_2}\alpha_M\alpha_X^2 \tag{4.49}$$

$$\Delta G = 2.3RT(\log K_{SMX_2} - 1/2\log K_{spMS} + 3/2\log\alpha_M - 1/2\log\alpha_S + 4\log\alpha_X - 2\log[X^-]), \tag{4.50}$$

where ΔG is the free energy change, ΔG_0 is the standard free energy, α_M is the second hydrolysis constant of metallic ions, K_{spMS} and K_{spMX_2} are solubility products of metallic sulfide and metallic ethyl xanthate, respectively; α_S and α_X are second constants of sulfide ions and xanthic acid, respectively; [X] is total concentration of xanthate. ΔG of reaction between ethyl xanthate and copper or iron sulfides calculated using the above relation as a function of pH is plotted in Fig. 4.42.

If the surface of sulfide minerals can be assumed to be in a stable oxidized state, the concentrations of metallic ions at the interface will increase. Assuming $[M] = 0.1$ mol/l and $[X] = 1.0 \times 10^{-4}$ mol/l and competition by OH^-, the critical pH for this can be estimated as follows:

$$M(OH)_2 + 2X^- = M(X)_2 + 2OH^- \tag{4.51}$$

$$K = [OH^-]^2/[X^-]^2 = K_{spM(OH)_2}/K_{spMX_2} \tag{4.52}$$

Thus, $pH_c = 10.8$ for PbS and 12.4 for CuS.

For $pH > pH_c$, from Eq. (4.52), ΔG is given by

$$\Delta G = -RT\ln K + 2RT\ln([OH^-]/[X^-]) \tag{4.53}$$

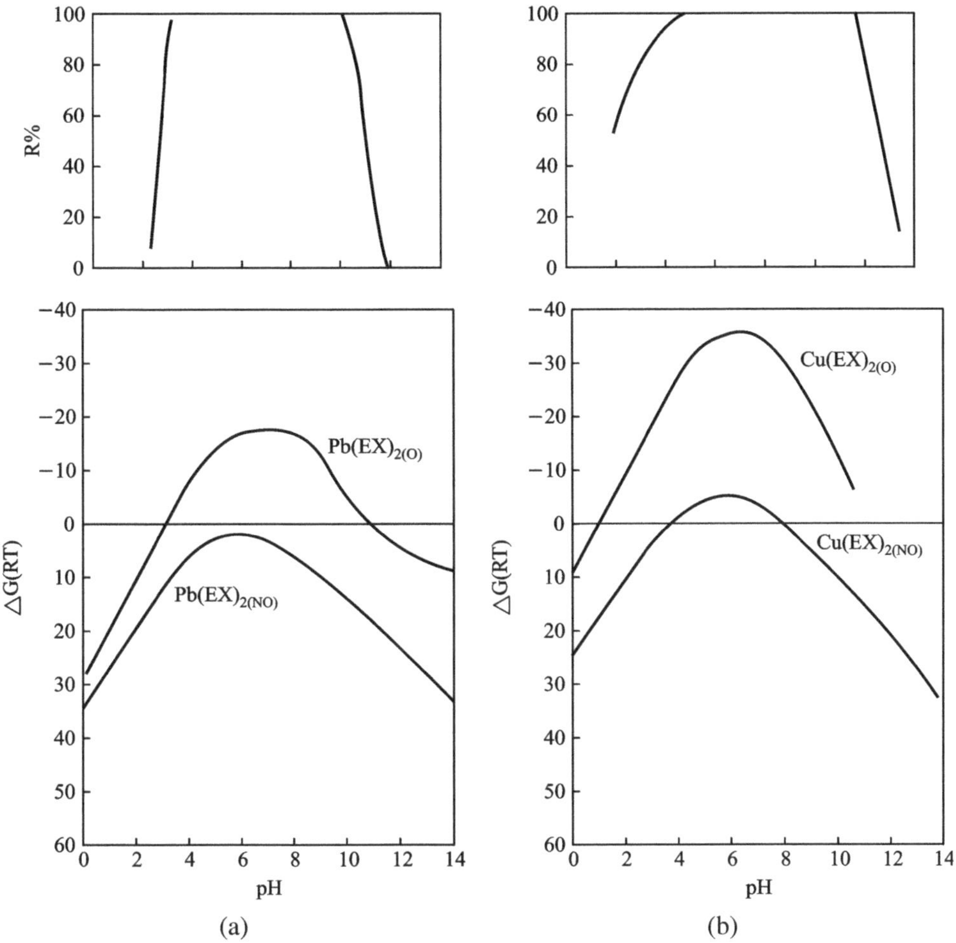

Fig. 4.43. ΔG–pH diagrams and flotation of: (a) Pb–EX system; (b) Cu–EX system (Hu, 1984).

For pH < pH_c,

$$[M^{2+}] = [M]/\alpha_M \tag{4.54}$$

Combining Eqs. (4.52) and (4.53),

$$\Delta G = 2.3RT(\log K_{SM(X)_2} + 2\log\alpha_M + 4\log\alpha_X - 2\log[X] - \log[M]) \tag{4.55}$$

ΔG calculated from the above equation as a function of pH for $Pb(EX)_2$ and $Cu(EX)_2$ are given in Fig. 4.43. The plots show that the pH value of maximum $-\Delta G$ corresponds to that of the maximum flotation. It is to be noted that reactions between xanthate and unoxidized copper or lead sulfides can be complex.

4.4.4.2. *Wolframite–collector system*

Consider reactions between metallic ions of wolframite and anionic collector ion to be as follows:

$$Mn^{2+} + 2A^- = MnA_2, \qquad K_{SMnA_2} = [Mn^{2+}][A^-]^2 \tag{4.56}$$

$$Fe^{2+} + 2A^- = FeA_2, \qquad K_{SFeA_2} = [Fe^{2+}][A^-]^2 \tag{4.57}$$

In addition, the second reaction of A^- with a proton is given below:

$$A^- + H^+ = HA, \qquad K^H = [HA]/[H^+][A^-] \tag{4.58}$$

Assume K_{sp} is the conditional solubility product. From Eqs. (4.56) and (4.58), standard free energy for the above reaction can be calculated as:

$$\Delta G^0_{Mn} = RT \ln K_{spMn} = RT \ln K_{spMnA_2}\alpha_{Mn}\alpha_A^2 \tag{4.59}$$

$$\Delta G^0_{Fe} = RT \ln K_{spFe} = RT \ln K_{spFeA_2}\alpha_{Fe}\alpha_A^2 \tag{4.60}$$

where K_{spMn} and K_{spFe} are conditional solubility products of Mn and Fe salts; α_{Mn} and α_{Fe} conditional hydrolysis constants for Mn^{2+} and Fe^{2+} respectively, and α_A the second proton constants of collector ions. ΔG–pH diagrams for Mn-Fe and collectors systems such as α-nitroso-β-naphthol, sodium oleate and octyl hydroxamate are shown in Figs. 4.44a–4.44c.

4.4.4.3. *Depression of galena with chromate salt*

The reaction between CrO_4^{2-} and Pb of galena is given below:

$$Pb^{2+} + CrO_4^{2-} = PbCrO_{4(s)}, \qquad K_{sp} = 10^{-13.8} \tag{4.61}$$

and reactions with the second proton are given as:

$$H^+ + CrO_4^{2-} = HCrO_4^-, \qquad K_1^H = 10^{6.49} \tag{4.62}$$

$$H^+ + HCrO_4^- = H_2CrO_4^-, \qquad K_2^H = 0.24 \tag{4.63}$$

The free energy change for the formation of $PbCrO_4$ is

$$\Delta G = RT \ln K_{sp}\alpha_{Pb}\alpha_{CrO_4^{2-}} - RT \ln[Pb^{2+}][CrO_4^{2-}], \tag{4.64}$$

where K_{sp} is the solubility product of $PbCrO_4$, α_{Pb} the second hydrolysis constant of lead ion and α_{CrO_4} the second proton constant.

If the galena surface is oxidized and assuming the total concentrations of lead ion at the interface to be 0.1 mol/l and total concentration of chromate ion to be C_T, ΔG of the reaction can be written as:

$$\Delta G = 2.3RT(\log K_{sp} + 2\log \alpha_{Pb} + 2\log \alpha_{CrO_4^{2-}} - \log C_T[Pb]) \tag{4.65}$$

The resultant ΔG–pH diagram is given in Fig. 4.45.

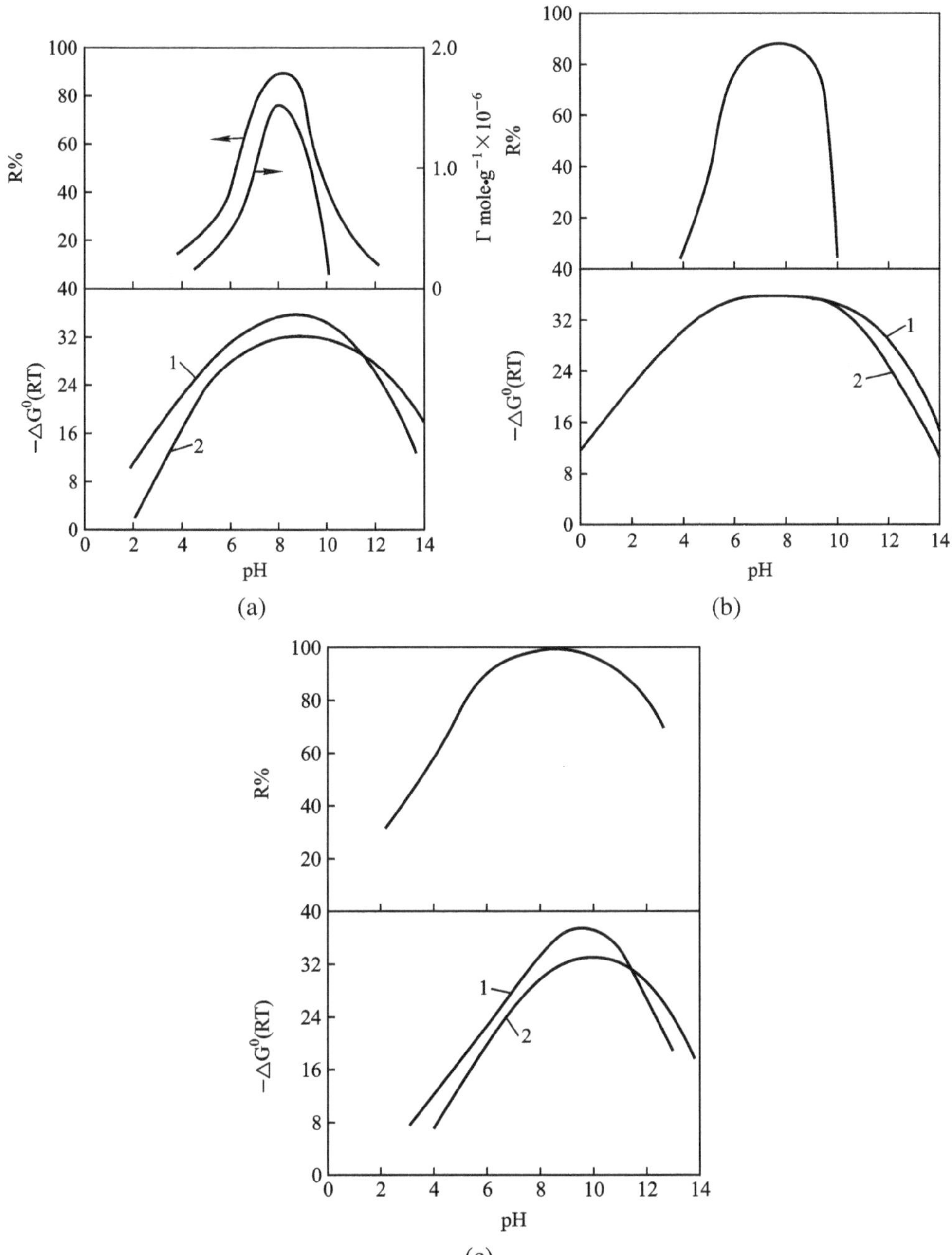

Fig. 4.44. ΔG–pH diagrams and wolframite flotation for Mn^{2+}-Fe^{2+}–reagent systems (Hu, 1984; Wang, 1986): (a) nitro-β-naphthol; (b) sodium oleate; (c) octyl hydroxamate. 1—MnL_2, 2—FeL_2.

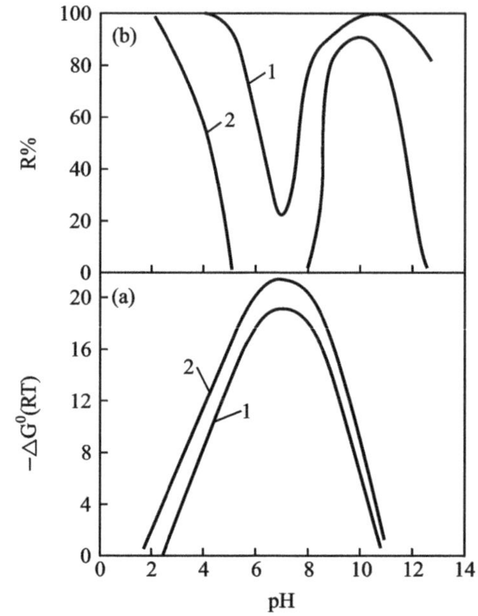

Fig. 4.45. ΔG–pH diagram of $PbCrO_4$ formation and galena flotation at (Shimoiizaka et al., 1976): 1—$[K_2CrO_4] = 500$ g/t; 2—$[K_2CrO_4] = 2000$ g/t, $[X^-] = 200$ g/t.

4.4.5. log β'_n–pH diagrams

A conditional stability constant, β'_n, for complexation reactions is used as a criterion for the reaction between metallic ions and reagents. The conditions for flotation of the minerals can then be predicted with the log β'_n–pH diagrams.

4.4.5.1. *Reactions between chelating agents and chrysocolla*

It has been shown that the combined use of 8-hydroxyquinoline, a chelating collector, and amyl xanthate can yield good flotation of chrysocolla at pH 7–10 (Mukai and Wakamatsu, 1975).

The reactions between Cu^{2+} and 8-hydroxyquinoline in aqueous solution are given as follows:

$$H^+ + L^- = HL, \qquad K_1^H = 10^{9.90} \tag{4.66}$$

$$H^+ + HL^- = H_2L, \qquad K_2^H = 10^{5.01} \tag{4.67}$$

$$Cu^{2+} + L^- = CuL^+, \qquad \beta_1 = 10^{15.0} \tag{4.68}$$

$$Cu^{2+} + 2L^- = CuL_2, \qquad \beta_2 = 10^{29.0} \tag{4.69}$$

$$Cu^{2+} + OH^- = CuOH^+, \qquad K_1 = 10^{6.3} \tag{4.70}$$

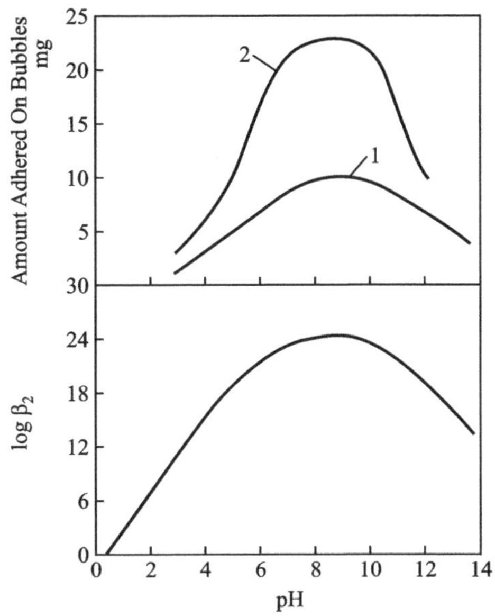

Fig. 4.46. $\log \beta_n'$–pH diagram of Cu^{2+}–8-hydroxyquinoline system and effect on chrysocolla flotation at hydroxyquinoline concentration: 1—10 mg/l; 2—30 mg/l: (Mukai and Wakamatsu, 1975).

$$Cu^{2+} + 2OH^- = Cu(OH)_2, \qquad K_2 = 10^{12.8} \tag{4.71}$$

$$Cu^{2+} + 3OH^- = Cu(OH)_3^-, \qquad K_3 = 10^{14.5} \tag{4.72}$$

$$Cu^{2+} + 4OH^- = Cu(OH)_4^{2-}, \qquad K_4 = 10^{16.4} \tag{4.73}$$

Assuming β_2' represents a conditional stability constant of CuL_2,

$$\beta_2' = [CuL_2]/[Cu^{2+}][L^-]^2 = 2\alpha_{CuL_2}/\alpha_{Cu}\alpha_L^2 \tag{4.74}$$

$$\alpha_L = 1 + K_1^H[H^+] + K_1^H K_2^H [H^+]^2 \tag{4.75}$$

$$\alpha_{Cu} = 1 + K_1[OH^-] + K_2[OH^-]^2 + K_3[OH^-]^3 + K_4[OH^-]^4 \tag{4.76}$$

where K_1^H and K_2^H are the first and second dissociation constants of 8-hydroxyquinoline, respectively; β_1' and β_2' stability constants of the first and second Cu–8-hydroxyquinoline complexes, respectively; K_1 to K_4 are stability constants for the first to fourth hydroxyl complexes, respectively; α_L is the second proton constant of 8-hydroxyquinoline and α_{Cu} second hydrolysis coefficient of copper ions and L represents ligand.

Based on the above equations, $\log \beta_n'$ values for various pH can be obtained, and the plot given in Fig. 4.46 shows the optimum pH region for the flotation of chrysocolla.

4.4.5.2. *Reactions between cyanide and copper, lead and zinc ions*

The reactions between cyanide and metallic ion are given below:

$$M^{2+} + nCN^- = M(CN)_n^{2-n} \tag{4.77}$$

$$\beta_n = [M(CN)_n^{2-n}]/\{[M^{2+}][CN^-]^n\} \tag{4.78}$$

$$H^+ + CN^- = HCN, \qquad K^H = 10^{9.21} \tag{4.79}$$

where β_n' is the stability constant of complex. The conditional stability constant is:

$$\beta_n' = \beta_n/\alpha_M \alpha_{CN}^n \tag{4.80}$$

$$\alpha_{CN} = 1 + K^H[H^+] \tag{4.81}$$

where α_M is the second hydrolysis coefficient of metallic ion, α_{CN} is the second proton constant and K^H is dissociation constant of hydrocyanic acid. The stability constant of cyanide with metallic ions is given as:

$$Cu^{2+}: \ \beta_4' = 10^{26.7}; \qquad Pb^{2+}: \ \beta_1 = 10^{2.4}$$

$$Zn^{2+}: \ \beta_4 = 10^{21.67}; \qquad Fe^{2+}: \ \beta_6 = 10^{35.4}$$

The conditional stability constant of various metallic ions can be obtained as a function of pH, using this equation along with the resultant $\log \beta_n'$–pH diagram shown Fig. 4.47. The diagram shows that Cu^{2+}, Fe^{2+}, Zn^{2+} ions can form stable complexes so that cyanide is a depressant for sulfides of copper, zinc and ferrous. However, lead ion cannot form stable cyanide complexes and thus does not depress the flotation of galena.

4.4.6. Species distribution diagrams

Due to a variety of possible solution reactions, flotation reagents exist in many forms such as undissociated molecules, ions, hydroxylated species and polymeric species under different solution conditions of pH and concentration. The fraction of species plotted as a function of the total concentration, pC, yields the species distribution diagram for the system. Such plots can be used to explain mechanisms by which reagents act in mineral flotation.

4.4.6.1. *Depression of flotation of wolframite and calcite using citric acid* (Wang and Hu, 1985)

Protonation reactions of citrate, the ligand L, can be written as follows:

$$H^+ + L^{3-} = HL^{2-}, \qquad K_1^H = 10^{6.40} \tag{4.82}$$

$$H^+ + HL^{2-} = H_2L^-, \qquad K_2^H = 10^{4.76} \tag{4.83}$$

$$H^+ + H_2L^- = H_3L, \qquad K_3^H = 10^{3.13} \tag{4.84}$$

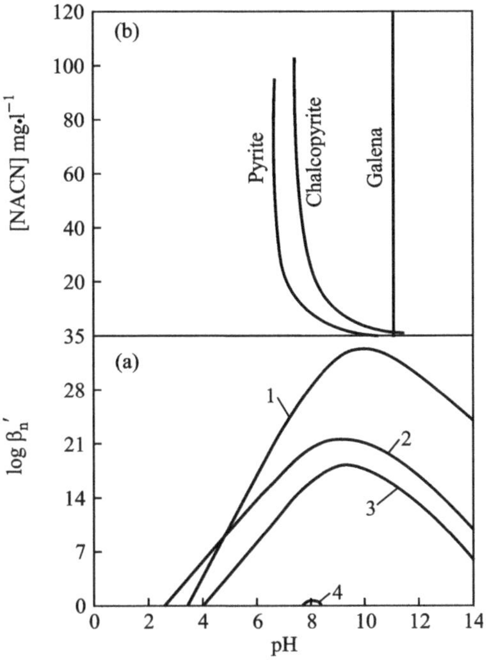

Fig. 4.47. $\log \beta_n'$–pH diagram of CN–Pb^{2+}, Cu^{2+}, Zn^{2+}, Fe^{2+} systems and NaCN concentrations required for depression of sulfide minerals, 1—$Fe(CN)_6^{4-}$; 2—$Cu(CN)_4^{2-}$; 3—$Zn(CN)_4^{2-}$; 4—$Pb(CN)^+$ (Sutherland and Wark, 1955).

Similarly the complexation of citrate with Ca can be written as follows:

$$Ca^{2+} + L^{3-} = CaL^-, \qquad K_1 = 10^{4.68} \tag{4.85}$$

$$Ca^{2+} + HL^{2-} = CaHL, \qquad K_2 = 10^{3.09} \tag{4.86}$$

$$Ca^{2+} + H_2L^- = CaH_2L^+, \qquad K_3 = 10^{1.10} \tag{4.87}$$

$$\begin{aligned} \alpha_{Ca} &= 1 + K_1[L] + K_2[L][H^+]K_1^H + K_3K_1^HK_2^H[H^+]^2[L] \\ &= 1 + C_T/\alpha_L\{K_1 + K_2K_1^H[H^+] + K_3K_1^HK_2^H[H^+]^2\} \end{aligned} \tag{4.88}$$

At pH $= 7.0$, $\alpha_L = 1.25$, then

$$\alpha_{Ca} = 1/(1 + 10^{4.59}C_T) \tag{4.89}$$

The distribution coefficients of various species are given by:

$$\phi_{Ca^{2+}} = 1/\alpha_{Ca} \times 100\% \tag{4.90}$$

$$\phi_{CaL^-} = K_1[L]\phi_{Ca^{2+}} \tag{4.91}$$

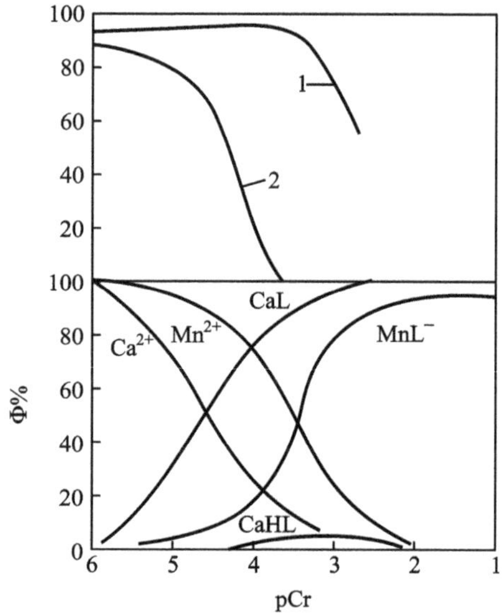

Fig. 4.48. Species distribution diagram of Ca^{2+}, Mn^{2+}–citric acid systems and flotation of calcite and wolframite using octyl hydroxamate (Wang and Hu, 1985).

$$\phi_{CaHL} = K_2 K_1^H [H^+][L]\phi_{Ca^{2+}} \tag{4.92}$$

$$\phi_{CaH_2L^+} = K_3 K_1^H K_2^H [H^+]^2[L]\phi_{Ca^{2+}} \tag{4.93}$$

where K_1^H, K_2^H and K_3^H are respectively the first, second and third stability constants of citric acid; K_1, K_2, and K_3 the first, second, and third stability constants of Ca–citric acid; α_{Ca} is the additional hydrolysis coefficient of Ca ion; α_L the additional proton constant of citric acid; C_T the total concentration of Ca ion; ϕ_{Ca} is the fraction of Ca ion and ϕ_{CaL}, ϕ_{CaHL} and ϕ_{CaH_2L} are respectively, fractions of first, second, and third Ca–citric acid complexes. The species distribution diagram of Ca–citric acid complexes obtained on the basis of the above treatment is given in Fig. 4.48.

The reaction of Mn^{2+} with citric acid is given by

$$Mn^{2+} + L^{3-} = MnL^-, \qquad K_1 = 10^{3.7} \tag{4.94}$$

$$Mn^{2+} + HL^{2-} = MnHL, \qquad K_2 = 10^{2.08} \tag{4.95}$$

The species distribution curves for these complexes are also shown in Fig. 4.48 along with the plots for citric acid depression of calcite and wolframite flotation. It can be seen that the depression of calcite by citric acid is higher than that of wolframite.

4.4.6.2. *ϕ–pC diagram of oxalic acid and depression with it*

Dissociation equilibria of oxalic acid are given below:

$$H_2C_2O_4 = H^+ + HC_2O_4^-, \qquad K_{a1} = 10^{-1.25} \tag{4.96}$$

$$HC_2O_4^- = H^+ + C_2O_4^{2-}, \qquad K_{a2} = 10^{-4.27} \tag{4.97}$$

and complexation reactions with Fe^{3+}, Fe^{2+} and Mn^{2+} ions are as follows:

$$Fe^{3+} + L^{2-} = FeL^+, \qquad K_1 = 10^{7.59} \tag{4.98}$$

$$Fe^{3+} + 2L^{2-} = FeL^-, \qquad \beta_2 = 10^{13.64} \tag{4.99}$$

$$Fe^{3+} + 3L^{2-} = FeL_3^{3-}, \qquad \beta_3 = 10^{18.49} \tag{4.100}$$

$$Mn^{2+} + L^{2-} = MnL, \qquad K_1' = 10^{3.2} \tag{4.101}$$

$$Mn^{2+} + 2L^{2-} = MnL^{2-}, \qquad \beta_2' = 10^{4.4} \tag{4.102}$$

$$Fe^{2+} + L^{2-} = FeL, \qquad K_1'' = 10^{3.05} \tag{4.103}$$

$$Fe^{2+} + 2L^{2-} = FeL^{2-}, \qquad \beta_2'' = 10^{5.15} \tag{4.104}$$

where K_{a1} and K_{a2} are respectively, the first and second dissociation constants of oxalic acid; K_1, β_2 and β_3 stability constants of Fe^{3+}–oxalic complexes in different forms; K_1' and β_2', K_1'' and β_2'' are respectively, stability constants of Mn^{2+}–oxalic complexes and Fe^{2+}–oxalic complexes. The species distribution diagrams for Fe^{3+}, Fe^{2+} and Mn^{2+} ions are shown in Fig. 4.49. It can be seen that the flotation of quartz activated by Fe^{3+} is depressed completely in the region where FeL_3^{3-} is formed. In the case of wolframite, the depression takes place in the regions where MnL_2^{2-}, FeL_2^{2-}, MnL and FeL complexes form.

4.5. DOMINANT SPECIES DIAGRAMS

In aqueous solutions, equilibria between reagents and metallic ions for various reactions such as hydrolysis, dissociation, precipitation and complexing will determine the concentrations of different chemical species. Under a given condition of concentration and pH, a given species will usually be dominant for the system. The dominant species diagram can be used to identify effective reagent species for the flotation process. It should however be kept in mind that even species that exist in smaller amounts can play a major role in determining the flotation.

As an example, solution equilibria for Na_2CO_3–$ZnCO_3$ system are given below:

$$ZnCO_{3(s)} = Zn^{2+} + CO_3^{2-}, \qquad K_{sp1} = 10^{-10.0} \tag{4.105}$$

$$Zn^{2+} + CO_3^{2-} = ZnCO_{3(aq)}, \qquad K_1 = 10^{5.3} \tag{4.106}$$

$$H^+ + CO_3^{2-} = HCO_3^-, \qquad K_1^H = 10^{10.33} \tag{4.107}$$

$$H^+ + HCO_3^- = H_2CO_3, \qquad K_2^H = 10^{6.35} \tag{4.108}$$

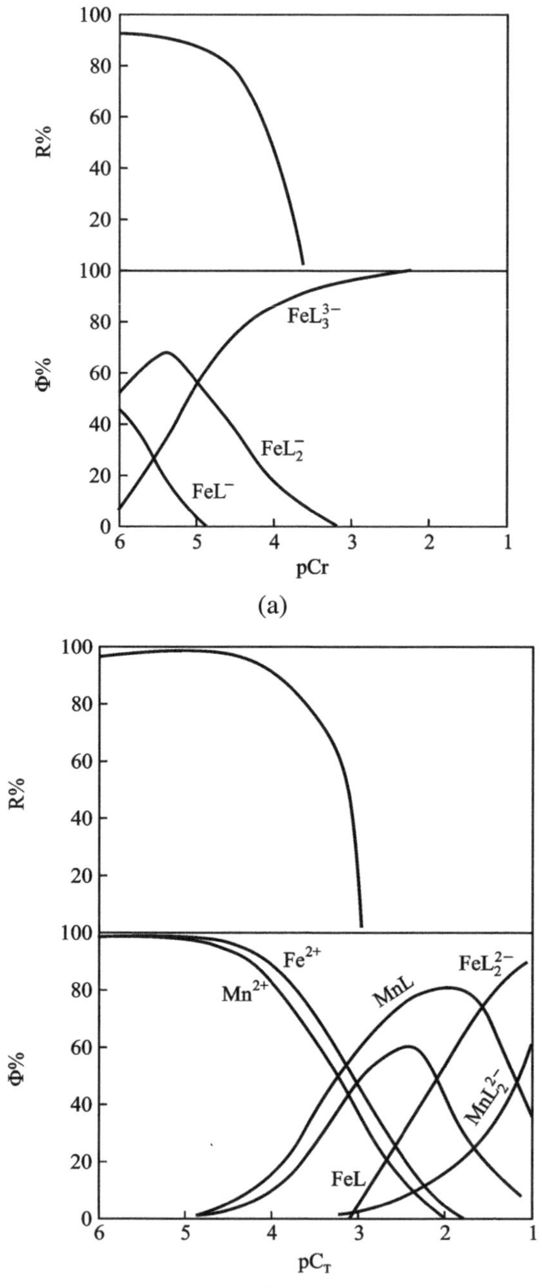

Fig. 4.49. Species distribution diagrams of metallic ions–citric acid systems and its depression for mineral flotation: (a) Fe(III)/quartz; (b) Mn(II), Fe(II)/wolframite (Wang and Bai, 1983).

$$Zn^{2+} + OH^- = ZnOH^+, \qquad \beta_1 = 10^{5.0} \tag{4.109}$$

$$Zn^{2+} + 2OH^- = Zn(OH)_{2(aq)}, \qquad \beta_2 = 10^{11.1} \tag{4.110}$$

$$Zn^{2+} + 3OH^- = Zn(OH)_3^-, \qquad \beta_3 = 10^{13.6} \tag{4.111}$$

$$Zn^{2+} + 4OH^- = Zn(OH)_4^{2-}, \qquad \beta_4 = 10^{14.8} \tag{4.112}$$

$$Zn(OH)_{2(s)} = Zn^{2+} + 2OH^-, \qquad K_{sp2} = 10^{-16.2} \tag{4.113}$$

where K_{sp1} is the solubility product of $ZnCO_{3(s)}$, K_1 equilibrium constant for formation of $ZnCO_{3(aq)}$, K_1^H and K_2^H are respectively stability constants of the first and second carbonic acids; β_1, β_2, β_3 and β_4 hydroxylation constants of the first to the fourth hydroxyl zinc complexes; K_{sp2} is solubility product of $Zn(OH)_{2(s)}$.

From the above equations, the demarcation lines for dominant species can be obtained as follows:

(a) Dividing line between $ZnCO_{3(s)}$ and $Zn(OH)_{2(s)}$

$$Zn(OH)_{2(s)} + CO_3^{2-} = ZnCO_{3(s)} + 2OH^- \tag{4.114}$$

$$\log[Na_2CO_3] = \log K_{sp1} - \log K_{sp2} - 2pK_w + \log \alpha_{CO_3} + 2pH, \tag{4.115}$$

where K_w is the ion product of $[OH^-]$ and $[H^+]$ of water, and α_{CO_3} is the additional protonation constant of carbonic acid.

(b) Demarcation line between Zn^{2+} and $ZnCO_3$. From Eq. (4.115),

$$\log[Na_2CO_3] = 1/2(\log K_{sp1} + \log \alpha_{CO_3} + \log \alpha_{Zn}), \tag{4.116}$$

where α_{Zn} is the additional hydrolysis coefficient of zinc ions.

(c) On the basis of Eq. (4.116), the following lines are obtained by

$$[Zn^{2+}] = [ZnOH^+], \qquad pH = 9.0 \tag{4.117}$$

$$[ZnOH^+] = [Zn(OH)_{2(l)}], \qquad pH = 7.9 \tag{4.118}$$

However, the dominant region for $ZnOH^+$ can not appear on the diagram and the relevant lines are in fact as follows:

$$[Zn^{2+}] = [Zn(OH)_{2(l)}], \qquad pH = 8.45 \tag{4.119}$$

$$[Zn(OH)_{2(l)}] = [Zn(OH)_3^-], \qquad pH = 11.5 \tag{4.120}$$

$$[Zn(OH)_3^-] = [Zn(OH)_4^{2-}], \qquad pH = 12.8 \tag{4.121}$$

The dominant species diagram for Na_2CO_3–$ZnCO_3$ system is given in Fig. 4.50a.

Conditions for $PbCO_3$ formation and the dominant species diagram are given as follows:

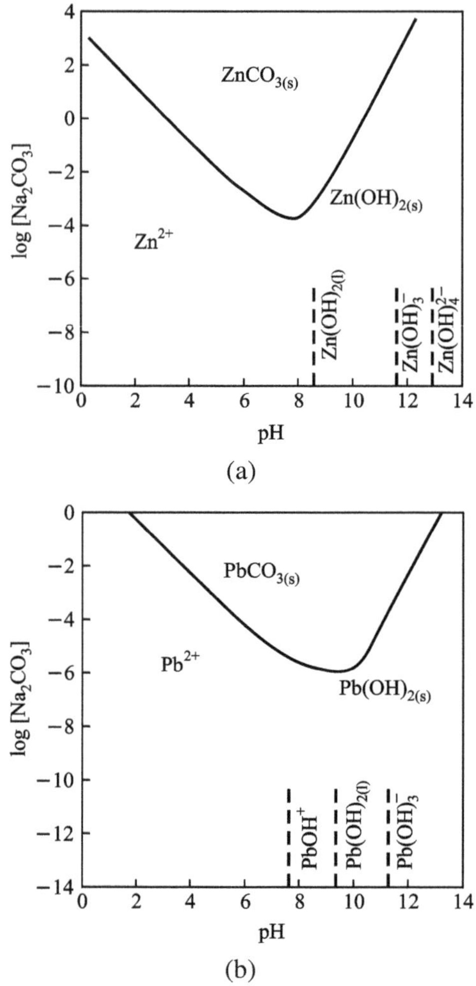

Fig. 4.50. Dominant species diagrams of: (a) Na_2CO_3–$ZnSO_4$–H_2O; (b) Na_2CO_3–Pb(II)–H_2O systems.

(a) $PbCO_3$–$Pb(OH)_2$ demarcation line

$$\log[Na_2CO_3] = \log K_{spPbCO_3} - \log K_{spPb(OH)_2} - 2pK_w + \log \alpha_{CO_3^{2-}} + 2pH \quad (4.122)$$

(b) Pb^{2+}–$PbCO_3$ line

$$\log[Na_2CO_3] = 1/2(\log K_{spPbCO_3} + \log \alpha_{Pb} + \log \alpha_{CO_3}) \quad (4.123)$$

(c) Dividing lines for various hydroxides are given by

$$[Pb^{2+}] = [PbOH^+] \qquad pH = 7.7 \quad (4.124)$$

$$[PbOH^+] = [Pb(OH)_{2(aq)}] \qquad pH = 9.3 \quad (4.125)$$

$$[Pb(OH)_{2(aq)}] = [Pb(OH)_3^-] \qquad pH = 11.0 \tag{4.126}$$

The dominant species diagram for Na_2CO_3–Pb^{2+} system is given in Fig. 4.50b. It can be seen from both diagrams that, in Na_2CO_3–$ZnCO_3$ aqueous systems, the pH region of $ZnCO_3$ formation is narrow but that of $PbCO_3$ formation is wide when the concentration of Na_2CO_3 is about 10^{-2} mol/l. It can also be seen that for Na_2CO_3–$ZnSO_4$ system, the role of Na_2CO_3 is to decrease the concentration of the lead in the pulp and to prevent activation when the dosage of Na_2CO_3 is not very high. In addition, it can increase the pH value and precipitate $Zn(OH)_{2(s)}$ causing depression of sphalerite flotation.

4.6. THREE-DIMENSIONAL (3-D) DIAGRAM

In flotation, often many variables will need to be investigated at the same time. For example, in the flotation of galena-sphalerite, excessive dosage of Na_2CO_3 in ethyl xanthate–$ZnSO_4$–Na_2CO_3 system will increase consumption of the xanthate. The reaction in this case is as given below:

$$PbX_2 + CO_3^{2-} = PbCO_3 + 2X^- \tag{4.127}$$

Thus, the concentration of xanthate needed for galena flotation will increase with sodium carbonate dosage with the corresponding relation given by:

$$\log[NaEX] = 1/2\log(K_{spPbX_2}/K_{spPbCO_3}) + 1/2\{\log\alpha_{CO_3} - \log\alpha_X + \log[Na_2CO_3]\}, \tag{4.128}$$

where K_{spPbX_2} and K_{spPbCO_3} are, respectively, solubility products of PbX_2 and $PbCO_3$, and α_{CO_3} and α_X are, respectively, second protonation constants of carbonic acid and ethyl xanthic acid.

Excessive amounts of $ZnSO_4$ will cause high consumption of xanthate with the corresponding relation given by

$$Zn^{2+} + 2X^- = ZnX_2 \qquad K_{spZnX_2} = [Zn^{2+}][X^-]^2 \tag{4.129}$$

$$\log[EX^-] = 1/2\log(K_{spZnX_2} - \log[Zn^{2+}]) \tag{4.130}$$

For galena and sphalerite flotation systems, the 3-D diagram for xanthate pH–Na_2CO_3–$ZnSO_4$–pH is shown in Fig. 4.51. It can be seen that the effect of $ZnSO_4$ on increasing the xanthate dosage is stronger than that of Na_2CO_3. Thus it is beneficial to increase the ratio of Na_2CO_3 to $ZnSO_4$ for this flotation system.

4.7. ELECTROCHEMICAL EQUILIBRIA

Sulfides as electronic conductors can act as donors of electrons and hence promote electrode reactions on mineral surfaces. Sulfide flotation systems constitute redox systems in

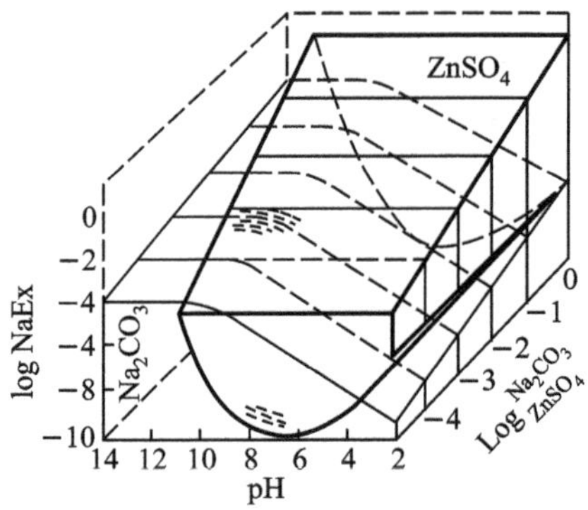

Fig. 4.51. 3-D diagram of depressant–xanthate–pH systems and selective flotation of galena and sphalerite (Wang, 1986).

which the sulfide mineral–reagent interactions are considered to take place by an electrochemical mechanism, in which the anodic process involving oxidation of mineral and collector is coupled with the cathodic oxygen reduction. The oxidation products of mineral and collector can be determined from the electrochemical equilibria involved (Trahar, 1983; Gardner and Woods, 1977; Tolley et al., 1996; Wang and Forssberg, 1996).

E_h–pH diagrams have become a standard method for illustrating equilibrium relationships between dissolved and solid species (Pourbaix, 1950). These relationships are represented with the activity of electrons (E_h) as the ordinate and the activity of protons (pH) as the abscissa. The oxidation-reduction potential or the redox potential (E_h) is a measure of the power of a solution to oxidize or reduce. Oxidation and reduction are essentially electron-transfer processes which are readily measured in terms of the electrical potential involved. All measurements are made in reference to the standard hydrogen electrode, the potential of which is assumed to be zero.

For a half-cell reaction,

$$aA + bB = cC + dD + ne^- \tag{4.131}$$

The relationship between the redox potential and a standard electrode potential is given by the Nernst equation:

$$E_h = E^0 + RT/nF \ln([C]^c[D]^d/[A]^a[B]^b), \tag{4.132}$$

where R is the gas constant (8.306 J/°C), T is temperature in K, F is the Faraday constant (96.40 kJ/°C), and n is the number of electrons involved in the redox reaction. E^0 is related to the standard free energy change by the relationship:

$$E^0 = \Delta G^0/nF \tag{4.133}$$

The standard free energy, G^0, of various dissolved metallic, sulfide, oxide and other species in aqueous solution are listed in the Appendix.

For a reaction at 25°C, $2.303RT/F = 0.059$. Then

$$E_h = E^0 + 0.059/n \log([C]^c[D]^d/[A]^a[B]^b) \quad (4.134)$$

4.7.1. Electrochemical equilibria of flotation agents

4.7.1.1. *E_h–pH diagrams of xanthate–H_2O systems*

Electrochemical equilibria of water are as follows:

For the upper E_h limit:

$$H_2O = 1/2O_2 + 2H^+ + 2e^-, \qquad E^0 = 1.23\text{ V}$$

$$E_h = 1.23 + 0.059/2 \log P_{O_2}^{1/2} a_{H^+}/a_{H_2O} \quad (4.135)$$

Under atmospheric conditions,

$$P_{O_2} = 2.1 \times 10^4\text{ Pa}, \qquad E_h = 1.22 - 0.059\text{pH} \quad (4.136)$$

For the lower E_h limit:

$$H_2 = 2H^+ + 2e^-, \qquad E^0 = 0.0\text{ V}$$

$$E_h = 0.059/2 \log a_{H^+}^2/P_{H_2} \quad (4.137)$$

The maximum possible partial pressure of hydrogen on the earth is 1.01×10^5 Pa, then $E_h = -0.059$pH.

The oxidation reactions of ethyl xanthate in solution are as follows:

$$2EX^- = EX_2 + 2e^-$$

$E^0 = -0.06$ V (literature value -0.037–0.15 V (Majima and Takeda, 1968; Finkelstein, 1967; Tolun and Kitchner, 1963; Harris, 1970; Kakovsky, 1957)) (4.138)

$$E_h = -0.06 - 0.059/2 \log[EX^-]/[EX_{2(l)}] \quad (4.139)$$

Assuming the activity of dixanthogen molecules to be unity, the xanthate–dixanthogen boundary is:

$$E_h = -0.06 - 0.059 \log[EX^-] \quad (4.140)$$

Dissociation reaction of xanthate is:

$$HEX = H^+ + EX^-$$

$$K_a = [H^+][EX^-]/[HEX] \quad \text{or} \quad [EX^-]/[HEX] = K_a/[H^+] \quad (4.141)$$

Then,

$$[EX^-] = [HEX] \quad \text{when} \quad K_a = [H^+] = 10^{-5} \quad (4.142)$$

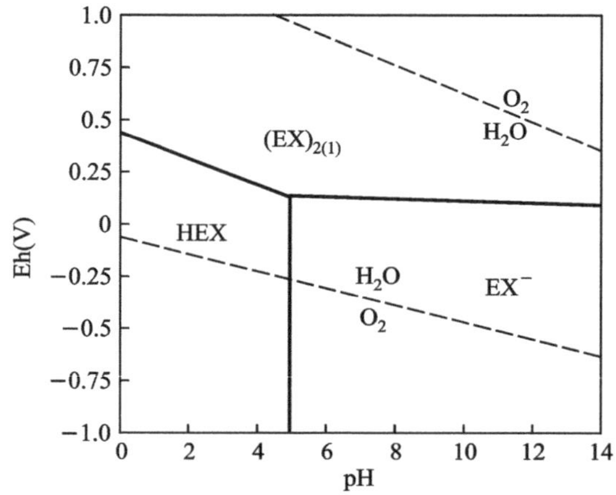

Fig. 4.52. E_h–pH diagram of ethyl-xanthate–H_2O system, $[EX^-] = 6.25 \times 10^{-4}$ mol/l.

Substituting into the above equation,

$$E_h = 0.235 - 0.059 \log[HEX] - 0.059pH, \tag{4.143}$$

which defines the boundary between HEX and EX. The E_h–pH diagram of xanthate at a concentration of 6.25×10^{-4} mol/l is shown in Fig. 4.52. It can be seen that dixanthogen is formed at the mineral surface when the rest or mixed potential is "anodic" to the equilibrium potential of the xanthate/dixanthogen couple, i.e. at pH $>$ 5 and $E_h > 0.13$ V or pH $<$ 5 and $E_h > 0.424 - 0.059$pH. This diagram also suggests the mechanism of sulfide flotation using ethyl xanthate: the mineral becomes hydrophobic due to the formation of the dixanthogen on the surface when its rest potential is larger than that of xanthate/dixanthogen couple; otherwise metallic xanthate salt is formed.

Table 4.2 lists the values of rest potential for a few minerals in potassium ethyl xanthate solutions (6.25×10^{-4} mol/l, pH 7) and infrared identifications of surface reaction products (Allison et al., 1972). Only those minerals such as chalcopyrite and pyrite have surface reaction product of dixanthogen.

4.7.1.2. *E_h–pH diagrams of diethyldithiocarbamate–H_2O systems*

Sodium diethyldithiocarbamate ($Et_2NCSSNa$) undergoes the following oxidation and dissociation reactions in water:

$$2Et_2NCSS^-_{(aq)} = (Et_2NCSS)_2 + 2e^-,$$

$$E^0 = -0.015 \text{ V (other literature values of } -0.068, -0.011 \text{ V (Finkelstein and Poling, 1977))} \tag{4.144}$$

$$E_h = -0.015 - 0.059 \log[Et_2NCSS^-] \tag{4.145}$$

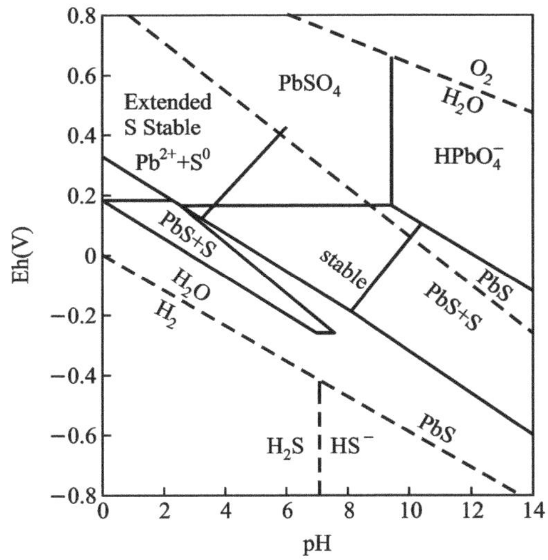

Fig. 4.53. E_h–pH diagram of galena–H_2O system with an energy barrier for SO_4^{2-} formation.

$$Et_2NCSSH = H^+ + Et_2NCSS^-, \qquad K_a = 10^{-5.6} \tag{4.146}$$

Then the boundary of Et_2NCSS molecules and ions is at pH 5.6.

Substituting Eq. (4.142) into Eq. (4.141), the boundary between Et_2NCSSH and $(Et_2NCSS)_2$ is:

$$E_h = 0.3154 - 0.059 \log[Et_2NCSSH] - 0.059\text{pH} \tag{4.147}$$

Finally, the boundary between Et_2NCSSH and Et_2NCSS^- is obtained from the reaction:

$$Et_2NCSSH = H^+ + Et_2NCSS^- \tag{4.148}$$

$$\frac{[Et_2NCSS^-]}{[Et_2NCSSH]} = \frac{K_a}{[H^+]} = \frac{10^{-5.6}}{[H^+]} \tag{4.149}$$

At the concentration of 5.8×10^{-4} mol/l, the E_h–pH diagram of diethyldithiocarbamate is shown in Fig. 4.53. At pH $>$ 5.6 and $E_h > 0.176$ eV or pH $<$ 5.6 and $E_h > 0.5064 - 0.059$pH, diethyldithiocarbamate can be oxidized in to disulfide and otherwise metallic diethyldithiocarbamate salts. Table 4.2 gives the values for the rest potential for a few minerals in sodium diethyldithiocarbamate solution (5.8×10^{-4} mol/l, pH 8) and the surface reaction products. Only pyrite has a higher potential than 0.176 V to form $(Et_2NCSS)_2$; other minerals should form $M(Et_2NCSS)_2$ on the surface.

4.7.2. Electrochemical equilibria of sulfide minerals

For PbS–S–O_2–CO_2–H_2O systems under open conditions, total dissolved sulfur activity is 0.1 and lead ion activity is 10^{-6}. At $P_{O_2} = 10^{-3.5}$, relevant electrochemical equilibria

Table 4.2
Rest potentials and surface reaction products in 6.25×10^{-4} mol/l KEX solutions

Minerals	Rest potential (V)	Surface product
Sphalerite	−0.15	Unidentified
Stibnite, Sb_2S_3	−0.125	Unidentified
Asenite	−0.12	Unidentified
Galena	0.06	MX
Chrysocolla	0.06	MX
Chalcocite	0.14	X2
Molybdenite	0.16	X2
Chalcopyrite	0.21	X2
Pyrite	0.22	X2
Arsenic pyrite	0.22	X2

are as follows:

$$(1)\ Pb^{2+} + HSO_4^- = PbSO_4 + H^+, \quad pH = 1.1 \tag{4.150}$$

$$(2)\ PbSO_4 + H_2O + CO_{2(g)} = PbCO_3 + SO_4^{2-} + 2H^+, \quad pH = 7.78 \tag{4.151}$$

$$(3)\ PbCO_3 + OH^- = HPbO_2^- + CO_{2(g)}, \quad pH = 13.3 \tag{4.152}$$

$$(4)\ HPbO_2^- = PbO_2 + H^+ + 2e^-, \quad E_h = 0.798 - 0.0295pH \tag{4.153}$$

$$(5)\ PbSO_4 + 2H_2O = PbO_2 + SO_4^{2-} + 4H^+ + 2e^-,$$
$$E_h = 1.6525 - 0.118pH \tag{4.154}$$

$$(6)\ PbCO_3 + H_2O = PbO_2 + CO_2 + 2H^+ + 2e^-, \quad E_h = 1.193 - 0.059pH \tag{4.155}$$

$$(7)\ PbS + HS^- + 10H_2O = HPbO_2^- + 2SO_4^{2-} + 20H^+ + 16e^-,$$
$$E_h = 0.688 - 0.0885pH \tag{4.156}$$

$$(8)\ Pb^{2+} + S + 4H_2O = HSO_4^- + 7H^+ + Pb^{2+} + 6e^-, \tag{4.157}$$
$$E_h = 0.3285 - 0.0688pH \tag{4.158}$$

$$(9)\ Pb^{2+} + S + 4H_2O = PbSO_4 + 8H^+ + 6e^-,$$
$$E_h = 0.3385 - 0.078671pH \tag{4.159}$$

$$(10)\ PbS + 4H_2O = PbSO_4 + 8H^+ + 8e^-, \quad E_h = 0.2983 - 0.059pH \tag{4.160}$$

$$(11)\ PbS + CO_{2(g)} + 5H_2O = PbCO_3 + SO_4^{2-} + 10H^+ + 8e^-,$$
$$E_h = 0.413 - 0.07375pH \tag{4.161}$$

$$(12)\ PbS + HS^- = PbS + S + H^+ + 2e^-, \quad E_h = -0.0365 - 0.0295pH \tag{4.162}$$

$$(13)\ PbS = Pb^{2+} + S + 2e^-, \quad E_h = 0.1772 \tag{4.163}$$

$$(14)\ (PbS + S) + 4H_2O = SO_4^{2-} + 8H^+ + 6e^-, \quad E_h = 0.3276 - 0.07867pH \tag{4.164}$$

$$(15)\ PbS + H_2S = (PbS + S) + 2H^+ + 2e^-, \quad E_h = 0.177 - 0.059pH \tag{4.165}$$

These boundaries are plotted in the E_h–pH diagram in Fig. 4.52. Under the condition of a 72 kcal/mol barrier for the formation of SO_4^{2-}, the extended stability zones of PbS or native sulfur are presented in Fig. 4.53. It can be seen from these figures that in the stability region of elemental sulfur, galena exhibits better collectorless floatability.

4.7.3. Electrochemical equilibria of minerals and reagents

4.7.3.1. *Flotation of galena and pyrite*

General mechanism of galena flotation using xanthate as collector is attributed to the formation of lead xanthate on the mineral surface:

$$PbS + 2X^- = PbX_2 + S + 2e^- \qquad (4.166)$$

$$2PbS + 4X^- + 3H_2O = 2PbX_2 + S_2O_3^{2-} + 6H^+ + 8e^-$$

$$E_h = 0.194 - 0.0295\log[X^-] + 0.00737\log[S_2O_3^{2-}] - 0.04425pH \qquad (4.166)$$

The equilibrium potential for reaction (4.166) in 6.25×10^{-5} mol/l xanthate solution is 0.12 V at pH 8. In the same xanthate solution and 10^{-6} mol/l thiosulfate solution, PbX_2 would be formed at −0.065 V at pH 8 and −0.19 V at pH 11, which are close to the lower E_h limits for flotation of galena in Fig. 4.54.

Electrochemical formation of dixanthogen on pyrite surface has been experimentally shown by several researchers (Finkelstein and Poling, 1977; Gardner and Woods, 1977; Usul and Tolun, 1974; Janetski et al., 1974). Fig. 4.55 shows that flotation of pyrite begins at 0.217 V at pH 8 and at 0.188 V at pH 9.2, respectively, for an initial KEX concentration of 2×10^{-5} and 6.25×10^{-5} mol/l as per the following reaction:

$$2EX^- = EX_2 + 2e^-, \qquad E_h = -0.06 - 0.059\log[X^-] \qquad (4.167)$$

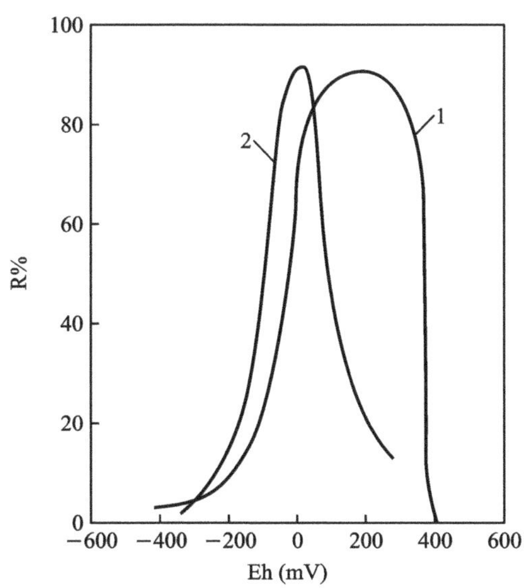

Fig. 4.54. Flotation of galena as a function of E_h: 1—pH = 8.0; 2—pH = 11.0 (Guy and Trahar, 1984; Trahar, 1983).

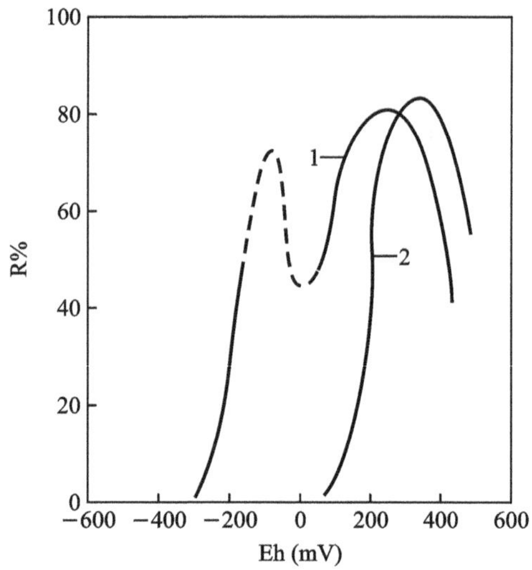

Fig. 4.55. Flotation of pyrite as a function of E_h: 1—pH = 8.0, [KEX] = 6.25 × 10^{-4} mol/l; 2—pH = 9.2, [KEX] = 2.0 × 10^{-4} mol/l (Trahar, 1983; Richardson and Walker, 1985).

A rapid decline in floatability of pyrite above 0.4 V is probably due to the decomposition of dixanthogen.

4.7.3.2. *Depression*

4.7.3.2.1. *Hydroxide ion* The oxidation reaction of pyrite is given below (Heyes and Trahar, 1984):

$$FeS_2 + 11H_2O = Fe(OH)_3 + 2SO_4^{2-} + 19H^+ + 15e^- \qquad (4.168)$$

Assuming a total sulfate concentration of 10^{-6} mol/l and a barrier to the formation of SO_4^{2-},

$$E_h = 0.7681 - 0.0747\text{pH} \qquad (4.169)$$

It is evident that E_h is decreased and the oxidation of pyrite itself increased as the pH is increased. At potentials lower than that of xanthate oxidation, the reaction (4.168) will take place preferentially and oxidation of xanthate will be suppressed. The critical depression pH value could be obtained from Eqs. (4.168) and (4.169):

$$\text{pH}_d = (0.828 + 0.059\log[X])/0.07473 \qquad (4.170)$$

At pH > pH_d, the process of xanthate oxidation into dixanthogen will be hindered and the flotation pyrite will be depressed as shown in Fig. 4.56 due to competitive adsorption between hydroxyl ions and xanthate ions.

Similarly, if the oxidation reaction of galena

$$2PbS + 7H_2O = 2Pb(OH)_2 + S_2O_3^{2-} + 10OH^- + 8e^- \qquad (4.171)$$

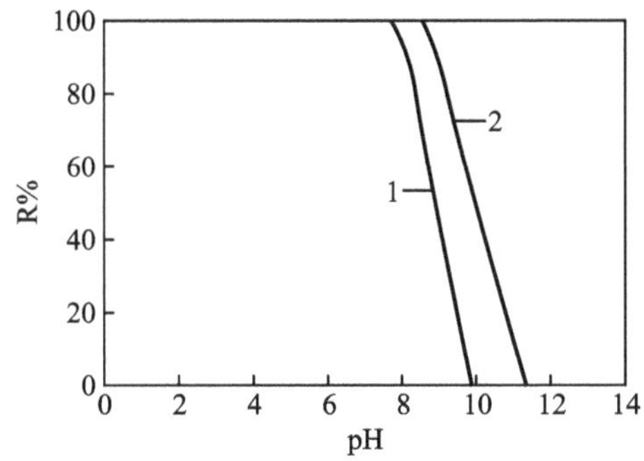

Fig. 4.56. Depression of pyrite flotation (Richard, 1976; Fuerstenau et al., 1968): 1—KOH, HCl; 2—CaO, [KEX] = 2.0×10^{-4} mol/l.

takes place before the formation of lead xanthate, the flotation of galena will be suppressed.

At a concentration of 10^{-6} mol/l thiosulfate, for reaction (4.171)

$$E_h = 0.59 - 0.0737\text{pH} \tag{4.172}$$

For the formation of lead xanthate (reaction (4.166))

$$E_h = 0.1498 - 0.0295 \log[X^-] - 0.04425\text{pH} \tag{4.173}$$

Therefore, the critical pH value for depression of galena can be obtained by substituting Eq. (4.172) into Eq. (4.173):

$$\text{pH}_d = 0.4402 + 0.0295 \log[X^-]/0.02945 \tag{4.174}$$

At pH > pH_d, oxidation of galena will lead to the formation of $Pb(OH)_2$ and $S_2O_3^{2-}$ at its surface and flotation of galena will be prevented.

4.7.3.2.2. *Hydrosulfide ions* The depression by hydrosulfide ions is based on critical concentration of HS^- ions relevant to xanthate oxidation, such that dixanthogen will not be formed and the mineral will not be rendered floatable.

The potential of HS^- ion oxidation to form SO_4^{2-} is given below:

$$E_h = 0.5953 - 0.066\text{pH} - 0.07375 \log[HS^-] \tag{4.175}$$

The critical pH value for hydrosulfide depression is calculated as follows:

$$\text{pH}_d = \{0.6553 + 0.059 \log[X^-] - 0.07375 \log[HS^-]\}/0.066 \tag{4.176}$$

At $[EX^-] = 1.56 \times 10^{-4}$ mol/l, the calculated pH_d value for depression of pyrite at different concentrations of HS^- is shown as a dashed line in Fig. 4.57, which is in agreement with the experimental data of Sutherland (solid line). At pH > pH_d, the oxidation of HS^- ions into SO_4^{2-} or $S_2O_3^{2-}$ will take place before X^- ions are oxidized to X_2 or begin to react with galena, and the mineral will not then be rendered hydrophobic.

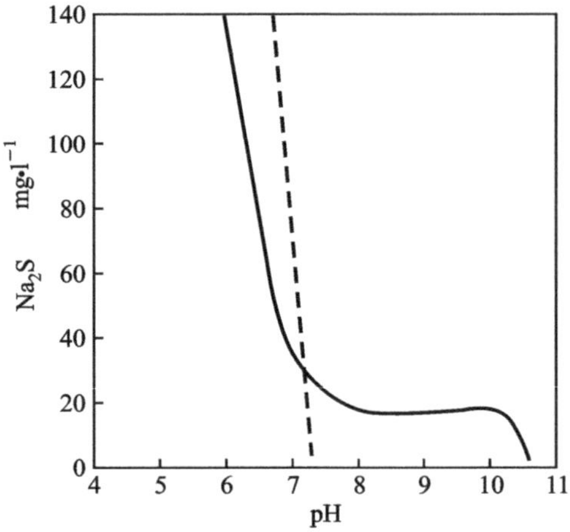

Fig. 4.57. Correlation of pH and Na_2S concentration required for preventing attachment of bubbles on pyrite (Sutherland and Wark, 1955).

4.7.3.2.3. *Cyanide* The depression of pyrite by cyanide is considered to invoke formation of an iron cyanide complex. The electrochemical reaction for the formation of a ferrous cyanide complex is shown below:

$$7Fe^{2+} + 18HCN = Fe_4[Fe(CN)_6]_3 + 18H^+ + 4e^- \quad (4.177)$$

The stability zone of $Fe_4[Fe(CN)_6]_3$ in Fig. 4.58 is in agreement with the condition of depression of pyrite with cyanide in Fig. 4.47. In fact, the presence of cyanide results in the inhibition of the electrochemical oxidation of the xanthate.

4.7.3.3. *Galena–xanthate systems*

Electrochemical reactions of galena with xanthate to form lead xanthate is given below:

$$2PbS + 3H_2O + 4X^- = 2PbX_2 + S_2O_3^{2-} + 6H^+ + 8e^-$$

$$E_h = 0.194 - 0.044pH + 0.0071\log[S_2O_3^{2-}] - 0.0295\log[EX^-] \quad (4.178)$$

The stability of PbX is defined by the following equation:

$$PbX_2 + 3OH^- = HPbO_2^- + 2EX^- + H_2O, \qquad K = 2 \times 10^{-3} \quad (4.179)$$

The decomposition potential of lead xanthate is determined by the relations:

$$PbX_2 + 2H_2O = HPbO_2^- + X_2 + 3H^+ + 2e^-$$

$$E_h = 1.225 - 0.0885pH + 0.0295\log[HPbO_2^-] \quad (4.180)$$

$$PbX_2 + 2H_2O = Pb(OH)_2 + X_2 + 2H^+ + 2e^-$$

$$E_h = 0.800 - 0.059pH \quad (4.181)$$

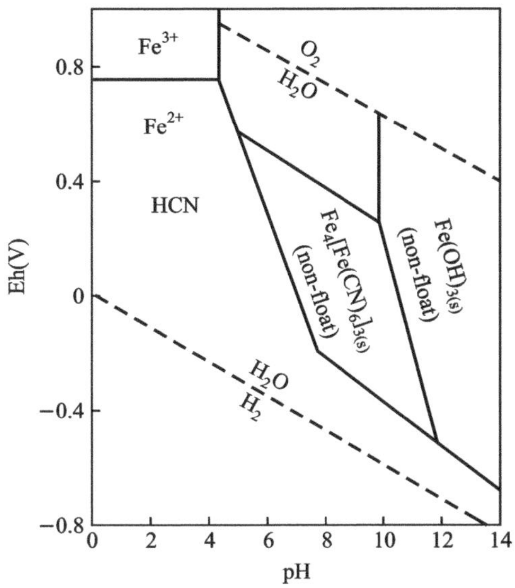

Fig. 4.58. E_h–pH diagram of pyrite–cyanide system ($[S]_T = 3.0 \times 10^{-4}$ mol/l, $Fe_T = 5 \times 10^{-5}$ mol/l).

The oxidation of galena leading to the formation of plumbite is

$$2PbS + 7H_2O = 2HPbO_2^- + S_2O_3^{2-} + 12H^+ + 8e^-$$

$$E_h = 0.841 - 0.0885pH + 0.015\log[HPbO_2^-] + 0.007\log[S_2O_3^{2-}] \quad (4.182)$$

$$Pb(OH)_2 = HPbO_2^- + H^+, \qquad K = 4.2 \times 10^{-5} \quad (4.183)$$

The oxidation of xanthate to form dixanthogen is given by

$$2EX^- = EX_2 + 2e^-, \qquad E_h = -0.081 - 0.059\log[EX^-] \quad (4.184)$$

The E_h–pH diagrams based on the above equations are plotted in Fig. 4.59 (total ionic strength kept at 10^{-4} mol/l). The region in which galena particle is contacted with gas bubble is that of the formation of the lead ethyl xanthate. This is the case for the flotation of galena in Fig. 4.54. The lead ethyl xanthate would be formed at pH 8 and at -0.06 V $< E_h < 0.36$ V or at pH 11 and -0.18 V $< E_h < 0.14$ V.

4.7.3.4. *Chalcocite–diethyldithiocarbamate (DTP) systems*

E_h–pH diagrams for chalcocite–DTP–water systems are shown in Fig. 4.60 based on the following reactions and E_h equations (Chander and Fuerstenau, 1975):

$$Cu_2S + 2DTP^- + H_2O = 2CuDTP + SO_4^{2-} + 8H^+ + 8e^-$$

$$E_h = 0.2785 - 0.0148\log[DTP^-] + 0.00788\log[SO_4^{2-}] - 0.059pH \quad (4.185)$$

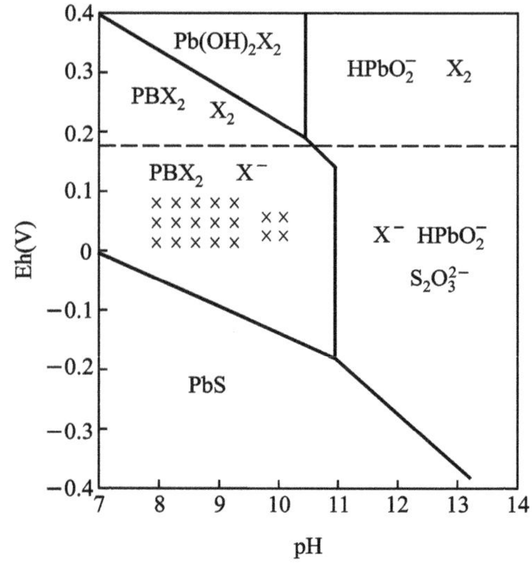

Fig. 4.59. E_h–pH diagram of PbS–EX–H_2O system ([EX] = 10^{-4} mol/l, [I] = 10^{-4} mol/l, × represents adhesion to bubbles) (Roperi and Tolun, 1969).

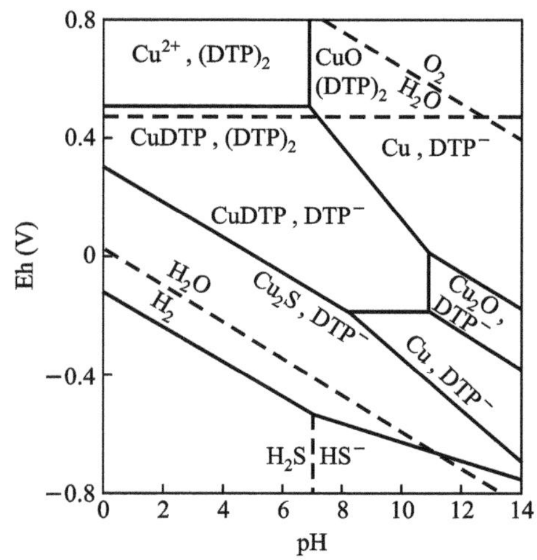

Fig. 4.60. E_h–pH diagram of DTP–Cu–H_2O system ([DTP$^-$] = 10^{-4} mol/l, assuming $[Cu^{2+}]_T = 10^{-6}$ mol/l, $[SO_4^{2-}]_T = 10^{-6}$ mol/l).

$$Cu_2S + 4H_2O = 2Cu + SO_4^{2-} + 8H^- + 6e^-$$

$$E_h = 0.506 + 0.00981 \log[SO_4^{2-}] - 0.0788\text{pH} \quad (4.186)$$

$$Cu + DTP^- = CuDTP + e^-, \qquad E_h = -0.416 - 0.059 \log[DTP^-] \quad (4.187)$$

$$2Cu + H_2O = Cu_2O + 2H^+ + 2e^-, \qquad E_h = 0.471 - 0.059\text{pH} \quad (4.188)$$

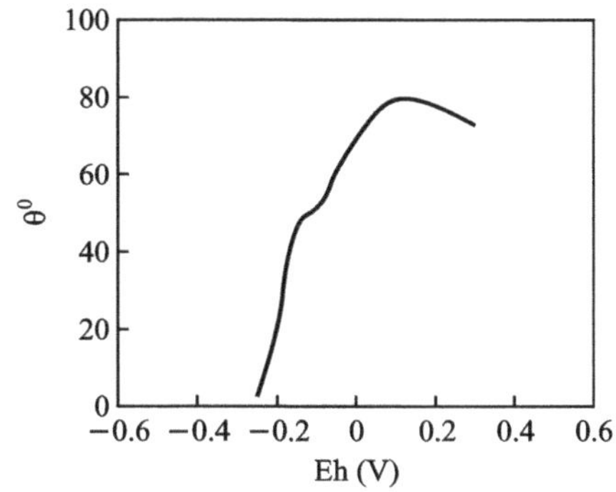

Fig. 4.61. Correlation of contact angle with E_h of chalcocite in KDTP solution ($[DTP^-] = 10^{-4}$ mol/l, 0.025 M $NaBO_3$) (Chander and Fuerstenau, 1975).

$$2CuDTP + H_2O = Cu_2O + 2DTP^- + 2H^+ + 2e^-$$

$$pH = (40.9 + 2.76 \log[DTP^-])/2.76 \tag{4.189}$$

$$Cu_2O + H_2O = 2CuO + 2H^+ + 2e^-, \qquad E_h = 0.669 - 0.059pH \tag{4.190}$$

$$CuDTP + H_2O = CuO + DTP^- + 2H^+ + e^-$$

$$E_h = 1.556 + 0.059 \log[DTP^-] - 0.1182pH \tag{4.191}$$

$$CuDTP = Cu^{2+} + DTP^- + e^-, \qquad E_h = 1.09 + 0.059 \log[Cu^{2+}][DTP^-] \tag{4.192}$$

$$CuO + 2H^+ = Cu^+ + H_2O, \qquad pH = 3.95 - 0.5 \log[Cu^{2+}] \tag{4.193}$$

$$2DTP^- = (DTP)_2 + 2e^-, \qquad E_h = 0.252 - 0.059 \log[DTP^-] \tag{4.194}$$

$$2Cu + H_2S = Cu_2S + 2H^+ + 2e^-, \qquad E_h = -0.305 - 0.0295 \log[H_2S] - 0.059pH \tag{4.195}$$

$$2Cu + HS^- = Cu_2S + H^+ + 2e^-, \qquad E_h = -0.512 - 0.0295 \log[HS^-] - 0.0295pH \tag{4.196}$$

The contact angle of chalcocite (Fig. 4.61) in KDTP solution begins to increase at E_h−0.2 V and to decrease at E_h 0.26 V. This is attributed to the formation of CuDTP at -0.18 V $< E_h <$ 0.22 V at pH 9.2. The metal–reagent complexes formed are responsible for the hydrophobicity.

The discussion in this section shows that E_h–pH diagrams can be used to identify stability regions for relevant species in mineral–reagent systems. The hydrophobic entity or flotation condition can also be estimated using this type of diagrams.

REFERENCES

Adamson, A.W., 1982. Physical Chemistry of Surfaces, 4th edn. John Wiley & Sons, New York.
Aiora, K.S., Turro, N.J., 1987. J. Polymer Science 25, 243.
Allison, S.A., et al., 1972. Metall. Trans. 3, 2613–2618.
Amankonah, J.O., Somasundaran, P., 1985. Colloids and Surf. 15, 335–353.
Ananthapadmanabhan, K.P., D. Eng. Sci. Thesis, Columbia University, New York, 1980.
Bahr, A., Clement, M., Surmatz, H., 1969. 8th Intemational Mineral Processing Congress, Leningrad, Paper S-11, p. 12.
Bisling, U., The mutual interaction of the minerals during flotation for example the fiotation of CaF_2 and $BaSO_4$, Dissertation, Bergakadamie Freiberg.
Bushell, C.H.G., 1958. Canad. Min. Metall. Bull. 51, 137.
Chandar, P., Somasundaran, P., Turro, N.J., 1987. Colloid & Interf. Sci. 117, 31.
Chandar, P., Somasundaran, P., Waterman, K.C., Turro, N.J., 1986. J. of Physical Chemistry 91, 150.
Chandar, P., Somasundaran, P., Waterman, K.C., Turro, N.J., 1987. Langmuir 3, 298.
Chander, S., Fuerstenau, D.W., 1974. Electroanalytical Chem. & Interfacial Electrochem. 56, 217–247.
Chander, S., Fuerstenau, D.W., 1975. 11th IMPC 1, 583–604.
Demas, J.N., 1983. Excited State Lifetime Measurements. Academic Press, New York.
du Rietz, C., Progress in Mineral Dressing, Stockholm, 1957, pp. 417–139.
du Rietz, C., 1975. 11th IMPC, pp. 375–403.
Finkelstein, N.P., 1967. Trans. IMM 76, 51–59.
Finkelstein, N.P., Poling, G.W., 1977. Miner. Sci. Eng. 9, 177–197.
Fuerstenau, D.W., et al., 1960. Trans. AIME 217, 111.
Fuerstenau, M.C., et al., 1968. Trans. AIME 241, 437.
Fuerstenau, M.C., Rice, A.D., Somasundaran, P., Fuerstenau, D.W., 1964. Trans. IMM (London), 73, 381.
Gardner, R.J., Woods, R., 1977. Aust. J. Chem. 4, 981–991.
Gaudin, A.M., Fuerstenau, D.W., 1955. Min. Eng. 7, 66.
Giles, C.H., McEwan, T.H., Smith, D., 1960. J. Chem. Soc., 3973.
Guy, P.J., Trahar, W.J., 1984. Int. J. Miner. Process. 12, 15–38.
Hanna, H.S., Somasundaran, P., 1976. In: Fuerstelnan, M.C. (Ed.), Flotation—A.M. Gaudin Memorial Volume, vol. 1. AIME, N.Y., pp. 197–272.
Hanna, H.S., Somasundaran, P., 1977. Physico-chemical aspects of adsorption at solid/liquid interfaces: Mahogany sulfonate/berea sandstone, kaolinte. In: Shah, D.O., Schecter, R.S. (Eds.), Improved Oil Recovery by Surfactant and Polymer Flooding. Academic Press, New York.
Harris, G.H., 1970. Encylopedia of Chemical Technology—Xanthates, vol. 22. Wiley, pp. 419–429.
Harwell, J.H., Bitting, D., 1987. Langmuir 3, 500.
Healy, T.W., James, R.O., Cooper, R., 1957. In: Advances in Chem. Series No. 79. Am. Chem. Soc., Washington, DC, p. 62.
Heyes, G.W., Trahar, W.J., 1984. Proc. Int. Symp. Electrochem. Minr. Met. Process, pp. 85–201.
Hollander, A.F., Somasundaran, P., Gryte, C.C., 1981. Adsorption of polyacrylamide and sulfonated polyacrylamide on Na-kaolinite. In: Tewari, P.H. (Ed.), Adsorption from Aqueous Solutions. Plenum Press.
Hu, Y., Study of flotation of wolframite with chelating agents and neutral oil, Master thesis, 1984, p. 9.
Janetski, N.D., et al., 1974. Int. J. Miner. Process. 1, 135–140.
Kakovsky, I.A., Proc. 2nd Int. Cong. Surf. Act., vol. IV, London, 1957, pp. 225–237.
Lakowicz, J.R., 1983. Principles of Fluorescence Spectroscopy. Plenum Press, New York.
Lawver, J.E., Wiegel, R.L., Snow, R.E., Huang, C.L., Proceedings of XIVth Int. Miner. Process. Congress, Session IV—Flotation, Paper IV-20, Toronto, 1982.
Leja, J., 1982. Surface Chemistry of Froth Flotation. Plenum Press, New York, p. 295.
Majima, H., Takeda, M., 1968. Trans. AIME/SME 241, 431–436.
Markovic, B., Somasundaran, P., Adsorption of Polyacrylic acid on porous and non-porous alumina, 205 ACS National Meeting, Denver, CO, Abstract, 1993, p. 83.
Morawetx, H., 1975. Macromolecules in Solution. John Wiley & Sons, New York.
Mukai, S., et al., 1962. J. of Japanese Society of Mining 9, 657.
Mukai, S., et al., 1972. J. of the Mineral Source (Japanese) 17, 324.

Mukai, S., Wakamatsu, T., 1975. 11th IMPC 1, 671–689.
Parfitt, G.D., Rochester, C.H., 1983. Adsorption of small molecules. In: Parfitt, G.D., Rochester, C.H. (Eds.), Adsorption from Solution at the Solid/Liquid Interface. Academic Press, New York, p. 3.
Parks, G.A., 1975. In: Riley, Skirrow (Eds.), Adsorption in Marine Environment, Chemical Oceanography. 2nd edn. Academic Press, p. 241.
Peck, A.S., Wardsworth (Eds.), 1965. In: Proc. 7th Int. Mineral Processing Congress, Arbiter, Gordon and Beach, p. 259.
Pourbaix, M., 1950. Thermodynamics of Dilute Aqueous Solutions. Arnold, London.
Ranby, B., ESR spectroscopy in polymer research, in: Rabek and Springer (Eds.), Berlin; New York: Springer-Verlag, 1977.
Richard, R.S., 1976. In: Fuerstenau, M.C. (Ed.), Flotation, A.M. Gaudin Memorial, vol. 1, pp. 458–484. Chap. 15.
Richardson, P.E., Walker, G.W., 1985. 15th IMPC, Cannes 2, 198–210.
Robert, O.F., et al., 1954. J. Phys. Chem. 58, 805.
Roperi, D., Tolun, R., 1969. Trans. IMM, C191–C197.
Shimoiizaka, J., et al., 1976. In: Fuerstenau, M.C. (Ed.), Flotation, A.M. Gaudin Memorial, vol. 1. AIME, New York. Chap. 13.
Siracusua, P., Somasundaran, P., 1986. Journal of Colloid and Interf. Sci. 114, 184.
Somasundaran, P., 1974. Trans. AIME 255, 64.
Somasundaran, P., 1976. Int. Journal of Mineral Processing 3, 35.
Somasundaran, P., et al., 1989. Langmuir 5, 215.
Somasundaran, P., Agar, G.E., 1967. J. Colloid Interf. Sci. 24, 333.
Somasundaran, P., Ananthapadmanabhan, K.P., 1985. Colloids & Surfaces 13, 151.
Somasundaran, P., Chandar, P., Turro, N.J., 1986. Colloids & Surfaces 20, 145.
Somasundaran, P., Fuerstenau, D.W., 1966. J. Phys. Chem. 70, 90.
Somasundaran, P., Fuerstenau, D.W., 1972. Trans. AIME 252, 275.
Somasundaran, P., Healy, T.W., Fuerstenau, D.W., 1964. J. Phys. Chem. 8, 3562.
Somasundaran, P., Kunjappu, J.T., 1989. Colloids & Surf. 37, 245.
Somasundaran, P., Lin, I.J., 1971. J. Colloid & Internt. Sci. 37, 731.
Sutherland, K.C., Wark, I.W., Principles of Flotation, Aust. Inst. Min. Metall., Melbourne, 1955, p. 127.
Sutherland, K.L., Wark, J.W., Principles of Flotation, Australasian Inst. Mining Met., Melbourne, 1955.
Taggart, A.F., et al., 1930. Amer. Inst. Min. Metal. Engr. Tech. Publ. 312, 3–33.
Thomas, J.K., The Chemistry of Excitation at Interfaces, American Chemical Society Monograph.
Tolley, W., Kotlyar, D., van Wagoner, R., 1996. Minerals Engineering 9 (6), 603–637.
Tolun, R., Kitchner, J.A., 1963. Trans. IMM 73, 313.
Trahar, W.J., 1983. The influence of pulp potential in sulfide flotation, principles of mineral flotation, The Wark Symposium, Aust. Inst. Min. Metall. Adlaide, Austalia, p. 117.
Turro, N.J., Cox, G.S., Paczkowski, M.A., 1985. Photochemistry in micelles. In: Topics in Current Chemistry, Photochemistry and Organic Synthesis. Boschke, F.L. (Ed.). Springer-Verlag, New York, vol. 129.
Usul, A.H., Tolun, R., 1974. Int. J. Miner. Process. 1, 135–140.
Wakamutsu, T., Fuerstenau, D.W., 1968. In: Weber Jr., W.J., Matijevic, E. (Eds.), Adsorption From Aqueous Solution. American Chemical Society, Washington, DC, p. 161.
Wang, D., 1986. Mineral Flotation and Reagent. Central-South University Press, Changsha, China, pp. 257–272.
Wang, Bai, 1983.
Wang, D., Hu, Y., 1985. Non-ferrous Metals 37 (4), 46–51.
Wang, D., Hu, Y., 1986. Mineral Flotation and Reagent. Central-South University Press, Changsha, China. p. 462.
Wang, D., Shibin, B., 1983. Non-ferrous Metals 37 (2), 47–53.
Wang, X.H., Forssberg, E., 1996. Mineral Engineering 9 (5), 527–546.
Wark, I.W., Cox, A.B., 1934. Trans. AIMME 112, 267.
Waterman, K.C., Turro, N.J., Chandar, P., Somasundaran, P., 1986. J. Phys. Chem. 90, 6829.
Wertz, J.H., Bolton, J.R., 1986. Electron Spin Resonance—Elementary Theory and Practical Apllications. Chapman and Hall, New York.
Willis, H.A., van der Maas, J.H., Miller, R.G.J. (Eds.), 1987. Laboratory Methods in Vibrational Spectroscopy. 3rd edn. John Wiley & Sons, New York.
Xiao, L., D. Eng. Sci. Thesis, Columbia University, 1990.

Chapter 5

Application of flotation agents and their structure–property relationships

5.1. CLASSIFICATION OF FLOTATION AGENTS

In general, flotation agents are classified according to their purpose as collectors, frothers, depressants, modifiers and flocculants. Structure of those molecules consists of polar and/or non-polar portions as shown in Fig. 5.1. Non-polar portion, which may be linear, branch or ring hydrocarbon, interacts very weakly with water molecules and therefore is termed hydrophobic group. the polar portion of the molecules can be classified into two types: one which strongly interacts with water molecules and termed *hydrophilic group*; the other which interacts with the surface of polar minerals and water molecules as well and termed minerophilic group. These three basic groups can be incorporated into the molecules to make various types of flotation agents.

With non-polar minerals such as talc, mica and coal, the particle surfaces can interact with the non-polar portion of the molecules by van der Waals force. In this case, hydrocarbon portions can function as mineralophilic groups. Except in the case of depressants or some flocculants, the non-polar portions of long hydrocarbon chains in flotation agents possess only the framework formed from C–H groups with linking to polar hydrophilic and minerophilic groups [4].

It is essential to maintain a proper balance among the three basic groups (hydrophobic, hydrophilic and minerophilic) in the molecules so as to get the desired collector, frother,

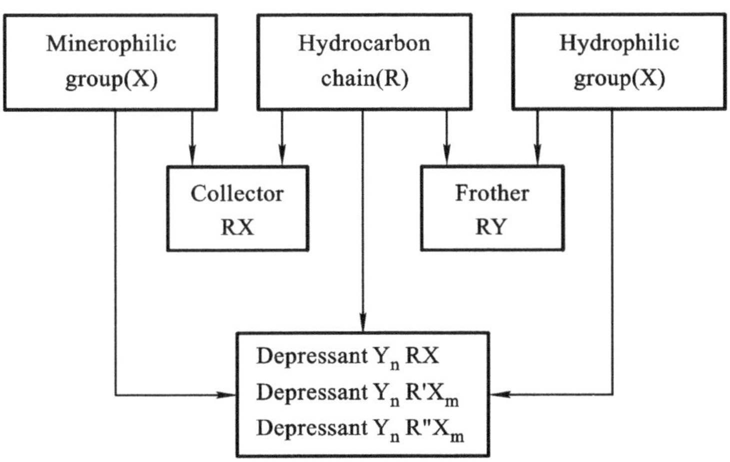

Fig. 5.1. Structure model of flotation reagents.

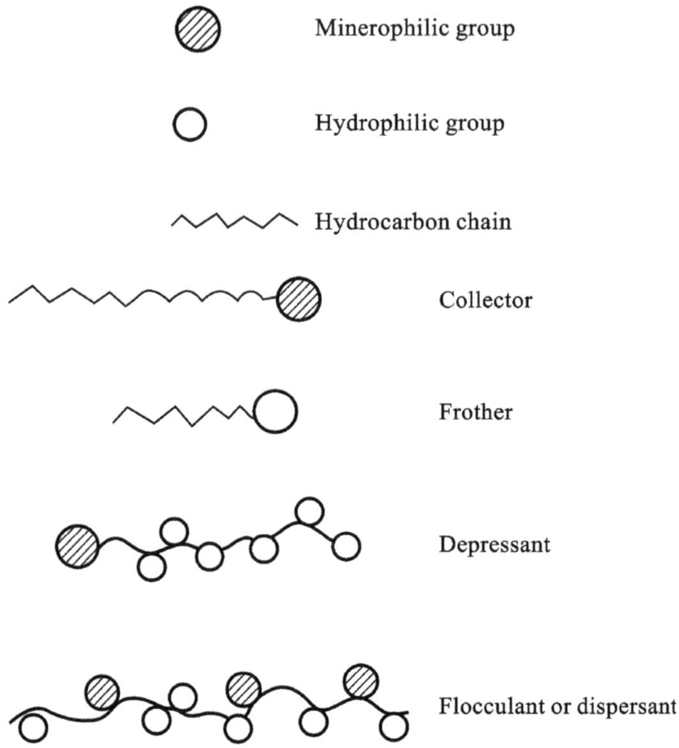

Fig. 5.2. Diagram of various flotation reagents formed from three elemental group.

depressant or flocculant properties. Various flotation agents are schematically shown in Fig. 5.2.

Each group of the molecule has an individual role in the flotation. Overall property of the molecule is a cumulative one of all the groups. A reagent molecule with various structural portions can therefore be divided into individual groups and its effect on the whole molecular property can be assessed. This facilitates evaluation of the structure–property relationships of flotation agents.

5.2. COLLECTORS—MINEROPHILIC AND NON-POLAR GROUPS

5.2.1. Structure of minerophilic groups

A collector is required to have at least two essential properties, namely adsorption on desired mineral particles and hydrophobization of their surfaces. Except for non-polar oils that are used sometimes as collectors for such naturally hydrophobic minerals like coal, the most common collector molecules possess minerophilic and hydrophobic portions in

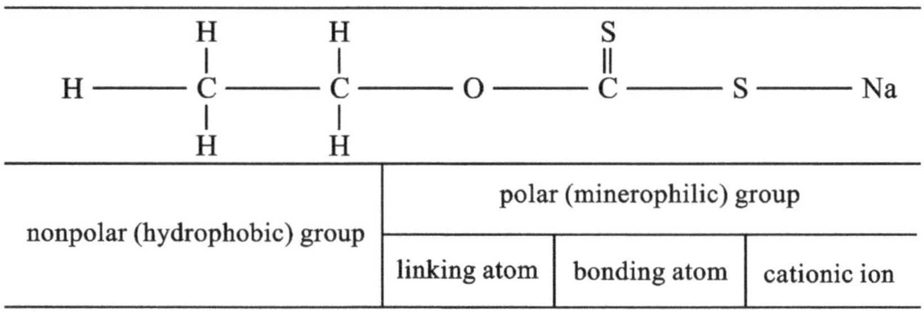

Fig. 5.3. Structure of minerophilic group of xanthate as collector for sulfides.

Table 5.1
pK_a values of various inorganic acids containing oxygen

pK_a	Acidity	Element				
		III	IV	V	VI	VII
−8	Very strong					$HClO_4$ (−8)
−3	Strong			HNO_3	H_2SO_4 (<0) H_2SeO_4 (<0)	$HClO_3$ (−3)
2	Sub-strong			HNO_2 (3.4) H_3AsO_4 (2.2)	H_2SO_3 (1.77) H_2SeO_3 (2.6)	$HClO_2$ (2.0)
7	Weak		H_2CO_3 (6.4)			HClO (7.5)
12	Very weak	H_3BO_3 (9.2) $Al(OH)_3$ (12.4)	H_2SiO_3 (10)	H_3AsO_3 (9.2)		
17	Very very weak		$Sn(OH)_2$ (14.4)		$Cr(OH)_2$ (16)	$Mn(OH)_2$ (19)

order to acquire the above properties. The minerophilic portion is a chemically functional group consisting of a bonding atom and linking atoms as shown in Fig. 5.3 [4].

Minerophilic groups determine the affinity of a collector to the mineral surface by means of mechanisms involving physical adsorption, chemisorption as well as chemical reactions due to differences in dissociation, solubility and polarity of the reagents. In addition, the cross-sectional area of the polar groups is usually larger than those of the non-polar groups, the geometric size of the minerophilic groups will also determine the cross-sectional area of the entire molecule which is a parameter that affects the flotation performance of a collector.

5.2.1.1. *Relationship between the structure of the minerophilic group and its adsorption on mineral surface*

Acids and salts composed of atoms of large electronegativity values such as halogens and oxygen usually exhibit a relatively high degree of dissociation. For certain acids containing oxygen, the more the number of oxygen atoms in a molecule, the higher is the degree of dissociation [5]. it is illustrated in Table 5.1 which shows the pK_a values of inorganic acids containing oxygen.

Dissociation degree of a reagent can be predicted using the atomic bonding parameter of polar group. Pauling [20] considered that the dissociation degree of inorganic acids containing oxygen depend on the charge of the central atom (m) and the number of oxygen atoms (n) without linking to any H atoms, and could be governed by the following equation:

$$pK_a = 7.5 - 10m + 5n \tag{5.1}$$

For organic bases such as amines, due to the electron donation effect and steric effect of alkyl groups, the alkalinity is in the order: primary amine > secondary amine > tertiary amine.

5.2.1.2. *Van der Waals' interactions between collector molecules and minerals*

Van der Waals force includes the molecular effects of orientation, induction and dispersion. The dispersion effect is generally the strongest and the induction is the weakest [16].

The relationship between the molecular structure and the van der Waals interaction can be discussed as follows:

(a) van der Waals interaction is determined by the dipole moment of a molecule. For a homologous series of organic compounds with the same polar group, the values of dipole moment and ionization potential are nearly the same, but value of polarity increases markedly with molecular weight. In other words, the dispersion and induction effects for homologous molecules will increase with molecular weight so that the adsorption depending on van der Waals force will also become stronger.

(b) Since the dipole moment is a vector property, it will increase with the degree of unsymmetrical distribution of polar and non-polar groups in a molecule. In general, the dipole moment of the molecules with a straight chain and para position substitution is larger than that of a molecule with branches and ortho position substitution. Thus, the van der Waals force of the former will be stronger than that of the latter.

5.2.1.3. *The polar group and hydrogen bonding*

In a molecule, the hydrogen atom linked to an atom (X) of high electronegativity can also be bonded with another atom (Y) of high electronegativity. This type of bonding is also called hydrogen bonding:

$$\text{–X–H} \cdots \text{Y–}$$

Thus, a molecule which has atoms of O, F, Cl, N or S of high electronegativity is small in size and can attract the electron from a hydrogen atom in another molecule and form hydrogen bonding.

Hydrogen bonding can result from interactions between reagents and H^+ and OH^- ions adsorbed on mineral surfaces or O, F and Cl atoms on minerals. In many cases, hydrogen bonding is the major force for reagent adsorption on minerals. For example, nonionic polymers such as polyacrylamide and polysaccharides are absorbed on minerals by hydrogen bonding.

Table 5.2
Minerophilic groups in collectors molecules

Carboxylate	R–C(=O)–OH
Sulfonate	$R–S(=O)_2–OH$
Phosphate	$R–P(=O)(–OH)_2$
Primary amine	$R–NH_2$
Tetra ammonium salt	$R(CH_3)_3–NX$
Arsenite	$R–As(=O)(–OH)_2$
Hydroxyl	R–OH
Hydroxamate	R–C(=O)–NH–OH
Mercaptan	R–SH
Thioether	R–S–R
Dithiocarbonate	R–O–C(=S)–S–Na
Dithiophosphate	$(R–O–)_2P(=S)–S–H$
Dithiocarbamate	$R_2N–C(=S)–S–Na$
Thionocarbamate	R–NH–C(=S)–O–R′

The energy of hydrogen bonding is of the order of 10^{11}–10^{12} ergs/mol which is smaller than that of chemical bonding. The hydrogen bonding energy between molecules of oleic acid is 2.1×10^{11} ergs/mol, but that between double bonds in the unsaturated chain of oleic acid is 1.36×10^{12} ergs/mol.

5.2.2. Minerophilic groups and chemical bonding

5.2.2.1. *Properties of bonding atoms*

Major minerophilic groups in collector molecules are listed in Table 5.2. The common bonding atoms in minerophilic groups are O, N, P and S which can be combined with C, P, As, Cl and Br elements to form many functional groups.

Interactions between the collector and the mineral surface are usually through these bonding atoms in the minerophilic groups and surface atoms (metallic ions) of the mineral crystal lattice. This will result in chemisorption or surface reactions discussed in Chapter 4. Characteristics of these bonding atoms as summarized by Nagaraj [3] are listed in Table 5.3.

5.2.2.2. *Properties of mineral constituent elements*

Considering the chemical affinity of atoms or ions and the resultant characteristic of minerals, constituent elements of minerals are classified in geochemistry into three categories as shown in Table 5.4. Affinity of bonding atoms in minerophilic groups to metallic ions on minerals are summarized in Table 5.5 [6].

5.2.2.3. *Bonding ability and specificity*

Solubility products of collector-metallic ion compounds (see Appendix C) suggest that the sulfide mineral collectors such as xanthates, mercaptans and thiophosphates containing sulfur bonding atom in the minerophilic group can form compounds of low solubility products with ions of elements with affinity for copper(II). They can not form insoluble

Table 5.3
The characteristics of bonding atoms

Characters	O	N	P	S
Electron configuration	$[He]1s^22s^22p^2$	$[He]1s^22s^22p^2$	$[He]1s^22s^23d^0$	$[He]1s^22s^23d^2$
Electronegativity	3.5	3.0	2.1	2.5
Valence electrons	2	5	5	2
Number of orbitals	4	4	4 + d	4 + d
Lone pairs	2	1	1	1
$p\pi$–$p\pi$	Strong	Strong	None	Poor
$d\pi$–$d\pi$ (back-bonding)	None	None	Moderate	Strong
Polarizability	Nil	Good	Good	Strong
Hydrogen bonding	Strong	Strong	None	Very weak
Bonds	More ionic	Less ionic	Covalent	Covalent
Steric accessibility	Low	Low	High	High

Table 5.4
Geochemical classification of mineral elements and interaction with reagents

Elements	Electron configuration of ions	Deposite character	Bonding atoms of reagent	Bonding character	Reaction with reagent
Elements affiniting rock (1st) Li, Be, Na, Mg, K, Ca, Al, Ba, Cl, Si	Inert gas type Li^+ $(1s^2)$He Be^{2+} $(1s^2)$He Na^+ $(1s^22s^22p^6)$Ne Mg^{2+} $(1s^22s^22p^6)$Ne K^+ $[(Ne)3s^23p^6]$Ar Ca^{2+} $[(Ne)3s^23p^6]$Ar	Later period of magmatism of magmaties stability in air	O, Cl	Ionic bond	Physical adsorption, partial chemisorption (for Ca, Ba)
Elements affiniting copper (2nd) Cu, Ag, Au, Zn, Cd, Hg, As, Sb, Bi, Sn, Pb, S, Se	Filled orbits d^{10} or $d^{10}s^2$	Hot solution stage components of polysulfides instability in air	S, Se	Covalent	Chemisorption and surface reaction
Element affiniting iron (3rd) Se, Ti, V, Cr, Mn, Fe, Co, Ni	Partially filled d or f orbits	In the earth's core and early period	N, C, P	Partially ionic	Physical adsorption and partial chemisorption

products with metallic ions of elements(I). The solubility products of compounds made up of ions of elements(III) and collectors with a sulfur bonding atom of sulfur depend upon the number of d electrons in the metallic ion shell and are generally higher than those of compounds made up of elements(II).

Table 5.5
Affinities of bonding atoms to metallic ions of minerals

Element(I) affiniting rock	O > N	N >> P >> As	O >> S > Se > Te	F >> Cl > Br > I
Element(II) affiniting copper	S > O	N << P > As	O << S, Se, Te	F < Cl < Br < I
Element(III) affiniting iron	O > N	(d-electron < 5)	N > O	(d-electron > 5)

Table 5.6
Δx of three kinds bonding

1st ions	Δx			2nd ions	Δx			3rd ions	Δx		
	O	N	S		O	N	S		O	N	S
Li	2.5	2.0	1.5	Cu	1.6	1.1	0.6	Ti	2.0	1.5	1.0
Na	2.6	2.1	1.6	Zn	1.9	1.4	1.9	Mn	2.0	1.5	1.0
K	2.7	2.2	1.7	Ag	1.6	1.1	1.6	Fe	1.7	1.2	1.0
Mg	2.3	1.8	1.3	Cd	1.8	1.3	0.8	Nb	1.9	1.4	0.7
Ca	2.5	2.0	1.5	Au	1.1	0.6	0.1	Cr	1.9	1.4	0.9
Ba	2.6	2.1	1.6	Hg	1.0	1.1	0.6				

Collectors for nonsulfide minerals with an oxygen as bonding atom and with long enough hydrocarbon chain length can form less soluble compounds with all three kinds of metallic ions, except K and Na ions. In this case the values of solubility products for various ions are only slightly different from those of each other.

Solubility products are usually used as a criterion of reactivity and specificity of collectors. In general, the less the solubility, the more is collectivity, but if the solubility is small enough to result in slight differences in solubility for various metallic ions, the selectivity will decrease. For separation of two minerals, it has been suggested that the ratio of the solubility products of compounds of two mineral ions (pL_1/pL_2) or the difference ($pL_1 - pL_2$) must be maintained at a suitable value [6].

The above discussion shows that the bonding of metallic ions with O atom is of ionic character and for the bonding of metallic ions(II) with S atom is covalent. Pauling has used electronegativity difference (Δx) as a criterion of ionic character, the ionic character being equal to 50% when $\Delta x = 1.7$. The Δx values for the three kinds of bonding are given in Table 5.6.

Collectors for nonsulfide minerals containing O atom, such as fatty acids and sulfonates, react with various metallic ions by ionic bond, and they have high solubility in water. The solubility becomes low only when the hydrocarbon chains in the molecules are long. In contrast, collectors for sulfide minerals, such as xanthates possessing only a short chain, the S bonding atom reacts with metallic ions to form covalent bonds with lower polarity, so the resultant products have low solubility in water and the solubility products are quite different for each metallic ions.

In other words, solubility products of short chain collectors for sulfides are determined mainly by the bonding character that reflects the properties of the metallic ions. With collectors of nonsulfide minerals, the lower solubility product is the result of "additive effects of long chain". In this case the bonding character is not closely related to the properties of metallic ions. Thus selectivity of collectors with S bonding atom is better than that of collectors with O bonding atom.

5.2.3. Induction effect between atoms in minerophilic groups

Polarity of unsymmetrical covalent bonds varies with the chemical structure around it. Due to differences in the electron-withdrawing power of the bonding atoms, electron clouds move toward the atom with higher electronegativity. This effect can be transmitted along the molecular chain, and this effect is called induction effect. The direction of the induction effect is determined by examining the electron-withdrawing power of an atom in relation to that of hydrogen atom. Thus if the electron-withdrawing power of an atom (X) in the molecule is larger than that of the H atom, the electron cloud moves to the X atom resulting in what is called negative induction (−I); otherwise it results in positive induction (+I) and represented as follows:

$R \rightarrow CH_2 \rightarrow X$

5.2.3.1. *Induction effect in molecules of thiocarbonates*

Among the derivatives of thiocarbonates, dithiocarbonates (xanthates) are widely used as collectors for sulfide minerals. In addition, monothiocarbonates and trithiocarbonates are also used. Bonding atoms and central atoms of all of those above thiocarbamates are the same, but the linking atoms are different:

(a) Monothiocarbonate: R–O–(O=)C←S←H(Me)
(b) Dithiocarbonate: R–O–(S=)C←S←H(Me)
(c) Trithiocarbonate: R–S–(S=)C←S←H(Me)

Electronegativity of the oxygen atom is higher than that of the sulfur atom. In comparing molecule (a) with (b) above, the −I effect of =O group is larger than that of the =S group, resulting in the lower electron density and bonding ability of S bonding atom of molecule (a) than that of the S in molecule (b). Similarly, in comparing (b) with (c), the bonding ability of trithiocarbonate to sulfide minerals is stronger than that of the dithiocarbonate, xanthate. As an example, the solubility products of lead thiocarbonates are given below:

- Lead ethylmonothiocarbonate $(C_2H_5OCOS)_2Pb$ $L = 1.5 \times 10^{-8}$
- Lead ethyldithiocarbonate $(C_2H_5OCS_2)_2Pb$ $L = 1.7 \times 10^{-19}$

Table 5.7
Properties of some thiocarbamates (σ^* is Taft constant)

Reagent	pK_a	σ^*	λ_{max}
(1) C_4H_9–O–C(=S)–NH–C(=O)–C_6H_5	7.45	+2.34	266
(2) C_4H_9–O–C(=S)–NH–CH–C_6H_5	11.5	+0.6	272
(3) *i*-C_3H_7–O–C(=S)=NH=CH=C_2H_5	11.53	–	–
(4) C_4H_9–O–C(=S)–NH–CH_2–C_6H_5	12.1	+0.215	246
(5) C_4H_9–O–C(=S)–NH–C_2H_5	12.7	−0.10	243
(6) C_4H_9–O–C(=S)–NH–C_2H_4OH	12.7	−0.10	245
(7) C_4H_9–O–C(=S)–NH–S–C(=O)–C_6H_5	–	–	–

5.2.3.2. *Induction effect in molecule of thionocarbamates*

Structure of the minerophilic group of thionocarbamate is shown as follows:

–O–C(=S)–NH–

Changes in the properties of a reagent with the addition of a donor atom or group can be explained by considering the induction effect. Table 5.7 shows structures and properties of some derivatives of thionocarbamates [4].

The electron density of the thiono-group (C=S) and its bonding ability to the proton will be decreased by the addition of an acceptor group such as –COC_6H_5, –C_6H_5 with a negative induction.

The collector will transfer a proton in acidic or alkaline medium:

R–O–C(=S)–NH–R′ $\rightleftarrows$ R–O–C(–SH)=N–R′

The dissociation degree and the acidity of the –SH group are increased by the addition of an acceptor groups. In contrast, The addition of a donor group such as –C_2H_5, results in +I effect. The ability of bonding to the proton increases and the acidity decreases.

Flotation performance of the collectors can also be predicted from data in Table 5.7 [4]. For collectors (1) and (7), due to introducing an acceptor, the bonding power of sulfur atom to iron ion is decreased markedly, but the bonding power to copper ion stays strong. Therefore, the selectivity of separation of copper mineral from pyrite is improved. However, the reactivity of the collector (2) to the two minerals is increased because of introducing a donor group with poor selectivity in flotation.

5.2.3.3. *Induction effect in molecules of thiophosphates*

Derivatives of thiophosphates include monothio and dithiophosphate with the following structures:

(a) (C_2H_5–O–)$_2$P(=O)–O–K
(b) (C_2H_5–O–)$_2$P(=S)–O–K
(c) (C_2H_5–O–)$_2$P(=S)–S–K

Solubility products of potassium salts of these reagents and the contact angles on galena in solution at a concentration of 1 mmol/l are given in Table 5.8.

Table 5.8
Solubility products of lead salts of thiophosphate derivatives and contact angles on galena (at 1 mmol/l)

Reagents	L_{PbAs}	Contact angle, θ	
		Ethyl	Propyl
(a)	Easily soluble	45	53
(b)	5.9×10^{-5}	51	67
(c)	2.2×10^{-10}	59	86

Molecule (a) is compared with (b), the bonding atoms and central atom are the same, but the difference in linking atoms, =S and =O, can result in different degree of hydrophobization of mineral surfaces ($\Delta\theta = 6°$). The molecules (b) and (c) possess identical linking atom but different bonding atoms, i.e. –S– and –O–, and this results in a significant difference in hydrophobization ($\Delta\theta = 18°$). The difference in the hydrophobization property of molecules between (a) and (c) is greater ($\Delta\theta = 24°$).

5.2.3.4. *Comparison of thiocarbonates with thiophosphates and thiocarbamates*

There are three types of reagents possessing the same bonding atoms but different central or linking atoms are shown below:

(a) xanthate: R–O–C(=S)–S–H(Me)
(b) dithiophosphate: $(R–O–)_2P(=S)–S–H(Me)$
(c) dithiocarbamate: $R_2N–C(=S)–S–H(Me)$

The effect of various atoms or groups is listed as below:

	(a)	(b)	(c)
Central atom	$X_c = 2.5$	$X_p = 2.1$	$X_c = 2.5$
	−I larger	−I smaller	−I larger
Linked atom	1 –O– group	2 –O– group	=N– group
	−I smaller	−I larger	−I smaller
Non-polar group	1 R-group	2 R-group	2 R-group
	−I smaller	+I larger	+I larger

In summary, the order of magnitude of solubility products of silver salts (in pL) and collective power for sulfide minerals are as follows: dithiocarbamate < xanthate < dithiophosphate as shown in Table 5.9.

5.2.3.5. *Collectors for nonsulfide minerals*

Dissociation degree and solubility products of metallic salts of common collectors are given below.

Reagent	Structure	pK_a	pL
Sulfonate	R–S(=O)$_2$–OH	~1.5	7–9
Phosphoric acid	R–P(=O)(–OH)$_2$	~2.6–2.9	
Arsenic acid	R–As(=O)(OH)$_2$	~3.7–4.7	
Carboxylic acid	R–C(=O)–OH	5	6–11

Table 5.9
Solubility products of silver alts with collectors

Non-polar group	Dithioncarbamate	Xanthate	Dithionphosphate
Ethyl	4.2×10^{-21}	4.4×10^{-21}	1.2×10^{-16}
Propyl	3.7×10^{-22}	2.1×10^{-19}	6.5×10^{-18}
Butyl	5.3×10^{-23}	4.2×10^{-20}	5.2×10^{-19}
Amyl	9.4×10^{-24}	1.8×10^{-20}	5.1×10^{-20}

5.2.4. Complexing (chelating) agents

As early as in the beginning of this century, chelating agents have been used as collectors in the flotation. For example, cupferron was used for flotation of oxidized copper minerals and 8-hydroxy quinoline for wolframite and lead–zinc minerals. Recently, chelating agents have been used as depressants by Schubert, Avotins and Nagaraj for the separation of fluorite from gangue minerals and molibdenite from copper sulfide minerals, respectively [3,13].

Complexing agents are classified as mono- and multi-dentates on the basis of the number of donors in their molecule. The multi-dentates can form ring structure with metallic ions and they are called chelating agents. The active atoms in the complexing agent are called donor or bonding atom or ligand atom and examples include mainly O, N, S and P atoms.

The common bonding atoms of chelating agents of the two-sets type includes S–S, N–N, O–O, S–N, S–O and O–N. Because most chemical groups are comprised of these atoms, many types of chelating compounds can be formed.

The amines commonly used in flotation processes possess one bonding atom in the molecule and hence can form only coordination compounds without a ring structure.

The stability of chelates is generally higher than that of coordination compounds, i.e. the chelating group has stronger bonding power with metallic ions. Therefore, chelating agents of lower molecular weight form less soluble chelates with metallic ions. On the other hand, complexing agents with mono-set generally produce soluble compounds and it was by increasing its molecular weight that insoluble complexes can be formed.

Structural characteristics of complexing agents are described as follows.

5.2.4.1. *Donor and acceptor atoms*

Various types of bonding of donors to Me^{2+} are described below:

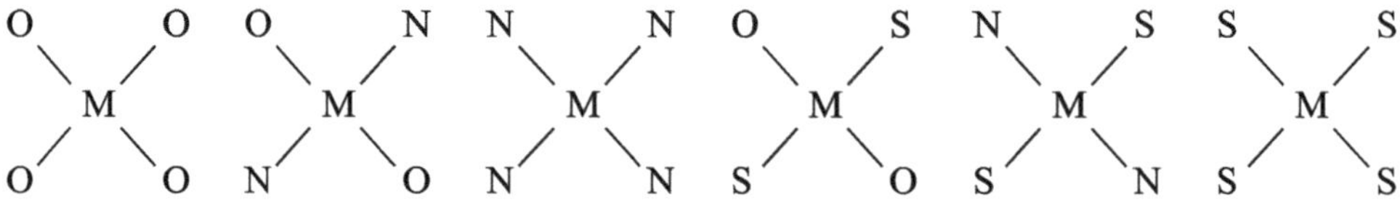

Bonding characteristics depend upon the coordination number and the steric structure:

(a) Ionic bonding: donors must be negatively charged and repel each other resulting in a tetrahedral configuration of four donors or an octahedral of six donors connected to the metallic ion in the center.
(b) Covalent bonding: the structure of common metallic compounds with covalent bonds can be classified as:

Be, Si, Cu, Zn, As, Sn, Pt, Cd, Mg	tetrahedron
Al, Cr, Mn, Fe, Co, Sn, Pb, Zn, As	octahedron
Cu, Ag, Au, Ni	tetragonal plane

The selectivity of chelating agents to mineral can be estimated from information involved in the donor class of and bonding:

Bonding characteristics	Minerals
Ionic	Minerals of metallic ions (1st)
Covalent	Minerals of metallic ions (2nd)
Interim	Minerals of metallic ions (3rd)

(c) Ring structure: for steric hindering effect of chelating rings, four-, five- and six-members are generally more stable. The five-member ring formed by saturated bonds and the six-member ring formed with two or more double bonds are also stable. Thus a favorable chelating group structure is that two bonding atoms are not directly linked and separated with less than three atoms forming a four, five or six-member ring. The chelating agents commonly used as flotation collectors are listed in Table 5.10.

5.2.5. Non-polar groups

5.2.5.1. *Role of non-polar groups*

Non-polar groups in collector molecules play the following three roles in determining the flotation performance:

(a) Hydrophobization of the mineral surface by hydrocarbon chains of the collector molecules adsorbed on the mineral surface.

Table 5.10
Chelating agents as collectors

Bonding atom	Polar group	Metallic ions	Application
N	–N=	II and III metallic ions	For nonsulfide minerals
N, N	-C-C= HO-N N-OH	Ni, Pd, etc.	For nickel and copper sulfide mineral
	–NH–C(=NH)–NH–	Zn, etc.	For ZnS
N, O	–N(–OH)–N=O	Cu, Fe, etc.	For copper mineral
	O=C=C=NO	Co, Zr, U, etc.	For wolframite
	O=C=C–NH–	Cu, Zr, Cd, Ta, etc.	For Ta, Nb mineral, wolframite
	–C(=O)–C(=NOH)–	Fe, Cd, Co, Ni, etc.	For oxidized copper minerals
	$HOOC–(CH_2)_4–NH_2$	Great part of metallic ions	For oxidized iron minerals and wolframite
S, S	–O–C(=S)–SH	II, III metallic ions	For sulfides
	$(–O–)_2P(=S)–SH$	Idem.	Idem.
	N–C(=S)–SH	Idem.	Idem.

(b) Associative chains interactions through van der Waals force between hydrocarbon chains, hydrogen bonding and electrostatic forces between double bonds and radicals in the non-polar portion.

Desorption of collector molecules from the mineral surface can occur only by overcoming not only affinity of the minerophilic group to the surface, but also the above forces associated with the chains. Thus the affinity of the collector to the mineral surface is also dependent on the structure of the non-polar groups.

(c) Non-polar groups can affect indirectly on the adsorption of polar groups through inductive, conjugative and steric effects. These effects are illustrated in Fig. 5.4.

According to the Traube's rule, the surface activity of homologous surfactants increases by 3.2 times when one $–CH_2–$ group is added to their hydrocarbon chain:

$$\Delta\sigma = a \ln(1 + bc), \tag{5.2}$$

where $\Delta\sigma$ is the change in surface tension; a is a parameter characterizing the polar group and it is a constant for the homologous series, and b is a parameter termed the surface active coefficient characterizing the non-polar group and for two neighboring homologous $b = 0.313$.

Flotation performance of homologous collectors reported by Wark et al. [19], it was found that contact angle of minerals treated with collector solutions depends only on the structure of the non-polar groups of the reagent, and not on any other portion. In other words, for various structures of non-polar groups, contact angle on sulfide minerals in-

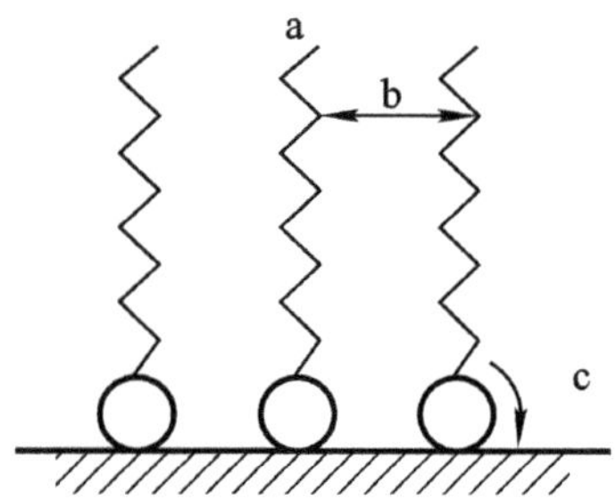

Fig. 5.4. Roles of non-polar group of collector molecule: a—hydrophobization, b—interaction between chains, c—influence of non-polar group on function of polar head (indirect influence on interaction between polar group and mineral).

Table 5.11
Contact angles of collectors for sulfide minerals

Reagents	Methyl	Ethyl	Butyl	Benzyl	Phenyl	Naphthatic
Dithiocarbamate	50	60	77	–	–	68
Mercaptan	–	60	74	71	70	68–71
Xanthate	50	60	74	72	–	71–75
Dithiophosphate	–	59	76	–	–	–

Table 5.12
Calculated values of contact angles (θ) of n-alkyl groups

Alkyl groups	Contact angle, θ^0	
	Experimental	Calculated*
C_3	68	68
C_4	74	74
C_5	80	79
C_6	90	90
C_7	94	94

*Using value of 68 for C_3.

creases only with the length of alkyl chains as shown in Table 5.11. Contact angles for various alkyl groups calculated from molecular parameters and given in Table 5.12.

Many flotation tests show that collectors possessing similar non-polar groups have different collecting properties. For example, ethyl xanthate is a good collector but acetic acid with the same ethyl group does not have any collecting property. It suggests that there is necessarily no definite parallel relationship between the above contact angle values and the flotation performance.

5.2.5.2. *Correlation of chain structure with its solubility and surface activity*

At the air/water interface adsorption is essentially an airophilic process of the hydrocarbon chains. Hence the change in free energy can be represented by the relationship for the vaporization process:

$$RT\ln(A_{n+1}/A_n) = 2950 \text{ J/mol} \tag{5.3}$$

Thus,

$$A_{n+1}/A_n = 3.29, \tag{5.4}$$

where A represents the surface activity of a chain containing one or more carbon atoms. Since the Traube's value is equal to 3.2, the corresponding relation for the solubility is given by:

$$RT\ln(S_n/S_{n+1}) = 3581 \text{ J/mol} \tag{5.5}$$

Therefore, a correlation between surface activity and solubility can be obtained as follows:

$$RT\ln(A_{n+1}/A_n) = RT(2950/3581)\ln(S_n/S_{n+1}) = RT\ln(S_n/S_{n+1})^{0.823}$$

or

$$A_n S_n^{0.823} = A_{n+1} S_{n+1}^{0.823} \tag{5.6}$$

The above equation shows that the increase in surface activity is inversely proportional to the decrease in solubility raised to the power of 0.823.

Free energies for various interactions between collectors and minerals are composed below:

Chain–chain interaction:	550–1200 cal/mol $-CH_2-$
Electrostatic interaction:	23 cal/mol per 1 mV
Hydrogen bonding:	a few kcal/mol
Chemisorption:	tens of kcal/mol

5.2.5.3. *Structure–property relationship of hydrocarbon chains*

5.2.5.3.1. *Normal saturated straight chains* Alkyl chains possess a saw-tooth planar structure with C–C–C bond angle of 109.5°. The distance between atoms linked to each other in the chain is 1.54 Å. Considering the diameter of carbon atom as 1.54 Å, the length of each $-CH_2-$ in the chain is 1.26 Å and the width is 4 Å. Since the chemical bond between two carbon atoms is formed by σ bonding of sp^3 hybridization electrons, the C–C axis can freely rotate.

It was found by Stambodliadis [12] that solubility product (pL) of salts of copper, iron and nickel with dialkyl dithiophosphates of various chain lengths is linearly related to the

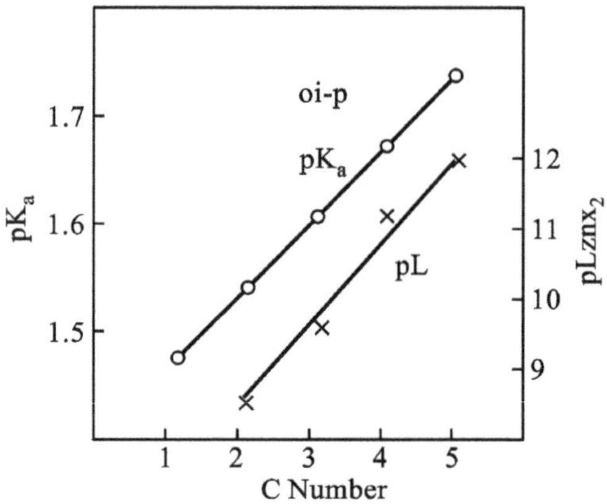

Fig. 5.5. pK_a and pL values as a function of the n values.

number of CH_2 groups (n):

$$\begin{aligned} &\text{Cu salt:} \quad pL = 10.77 + 12.58 \log n \\ &\text{Fe salt:} \quad pL = -10.07 + 39.6 \log n \\ &\text{Ni salt:} \quad pL = -20.54 + 40.96 \log n \end{aligned} \tag{5.7}$$

The solubility products (pL) and the dissociation constants (pK_a) of zinc-xanthates as a function of chain length (n) are illustrated in Fig. 5.5. The recovery of galena flotation using xanthates and fatty acids with various chain lengths is shown in Fig. 5.6 [1].

In the case of quartz flotation using primary amines and quaternary ammonium salts, as collectors, Somasundaran [24] found that the concentration ($\log C$) of collectors at the critical flotation condition is linearly related to n values as shown in Fig. 5.7. In addition, the concentration ($\log C$) at which the zeta-potential of quartz in alkylamine solutions is equal to zero was found to have similar linear relationship with n values (Fig. 5.8).

5.2.5.3.2. *Iso-alkyl chains* Flotation performance of collectors with an iso-structure chain is different from that of collectors with a linear structure. This is summarized below:

(a) Branched chains hinder proximately between the chains and these weaken the van der Waals force and results in low melt point, high solubility and CMC.
(b) Branched chain near to a polar group may possess steric hindrance between the polar group and the mineral surface.
(c) The effect of methyl group in an iso-structure close to the polar end and with inductive effect donating electrons from the methyl group can increase electron density of the bonding atom resulting in a strong bonding and have more collective power. For example, in the case of normal propyl xanthates and iso-propyl xanthate with the following structure:

normal: $CH_3–CH_2–CH_2–OCSSH$

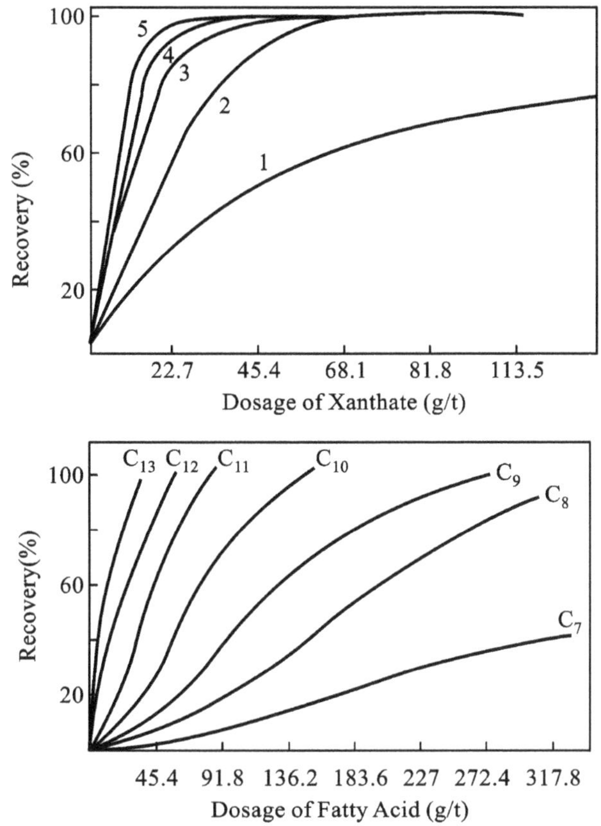

Fig. 5.6. The performance of xanthate and fatty acids with various chains for galena flotation: 1—methyl; 2—ethyl; 3—propyl; 4—butyl; 5—amyl.

iso-: $(CH_3-)_2CH-OCSSH$

Iso xanthates have two methyl groups linked to the carbon atom of the –CH in iso-propyl xanthate, hence the performance of iso-propyl xanthate is better than that of normal one.

(d) Chains with iso-structure possess larger cross-sectional area and thus yield a larger hydrophobic area when adsorbed on the mineral surface. This results in weak selectivity and strong frothing.

In summary, for reagents with short chains among the above factors, (c) is a major one and thus reagents with iso-structure are better collectors than the normal ones. However, for long chains, (a) and (b) factors are also important. The solubility of iso-structure can be increased as desired by decreasing the hydrophobic chains.

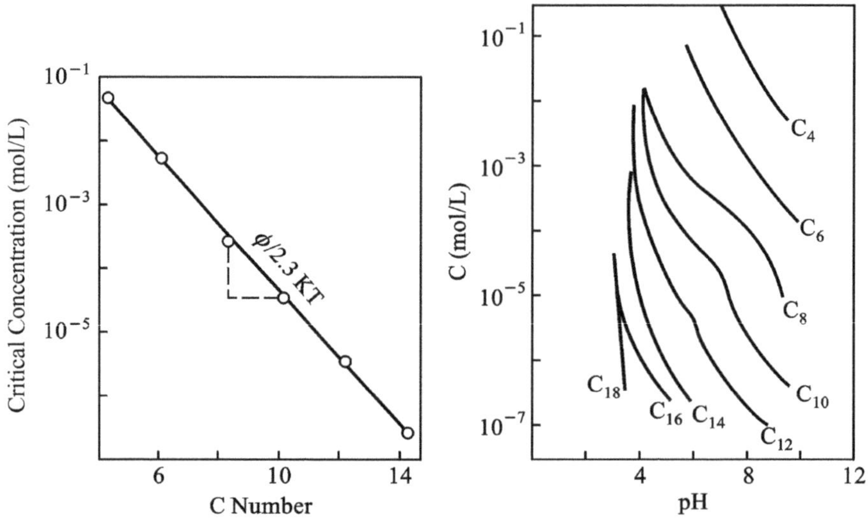

Fig. 5.7. Critical concentration of amines in quartz flotation (by P. Somasundaran, D.W. Fuerstenau).

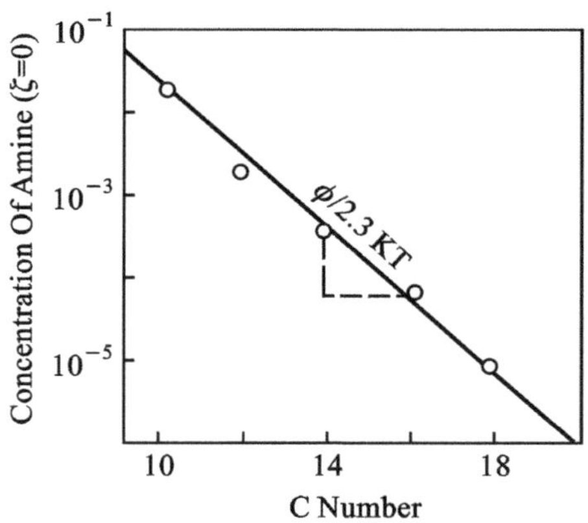

Fig. 5.8. Concentration of amines as a function of n values for quartz flotation (by P. Somasundaran).

5.2.5.4. *Unsaturated chains*

In comparison with saturated chains containing the same number of carbon atoms, organic compounds possessing unsaturated chains exhibit a lower melting point, higher solubility, CMC, chemical activity, and larger chain cross-sectional area. This is shown in Table 5.13. Four factors that affect the solubility products of reagents with unsaturated chains are:

(a) The number of double bonds.
(b) Position of double bonds in the chain.
(c) Steric structure in trans and cis forms of the chains.

Table 5.13
Properties of unsaturated fatty acids

Acid	Formula	CMC of K or Na salt	K_3P CaA_2 (20°C)	Melt point (°C)	Cross-sectional area of chain (λ_2)
Stearic acid	$CH_3(CH_2)_{15}COOH$	0.0045*	17.4	65	24.4
Oleic acid	$CH_3(CH_2)_7CHCH(CH_2)_7COOH$	0.0012*	12.4	16.3	56.6
Linoleic acid	$CH_3(CH_2)_4CHCHCH_2CHCH(CH_2)_2COOH$	0.15*	12.4	–	59.9
Linolenic acid	$CH_3(CH_2CHCH)_3CH_2(CH_2)_6COOH$	0.20**	12.2	12.5	68.2

*For K salt.
**For Na salt.

(d) Conjugated and separated double bonds.

5.2.5.5. *Aromatic groups*

A typical non-polar group with aromatic structure present in collector molecules is benzene in which six carbon atoms and six hydrogen atoms are formed into a closed conjugation system. Although the benzene ring contains six carbon atoms, its bonding and steric characteristics are quite different from those of molecules with straight alkyl chains containing the same number of carbon atoms. A benzene ring has a length of 2.84 Å and a width of 1.85 Å, which is analogous to ethyl group with higher hydrophilicity resulting from its conjugative structure.

It has been reported that the effect of adding phenyl group in non-polar portion on CMC value is analogous to the addition of 3.5 $-CH_2-$ groups. Solubilities of zinc-xanthate salts with phenyl group are given below along with those of corresponding alkyl xanthates:

Zn benzyl xanthate $(C_6H_5CH_2OCSS)_2Zn$	$S = 1.5 \times 10^{-4}$
Zn *n*-butyl xanthate $(C_4H_9OCSS)_2Zn$	$S = 2.1 \times 10^{-4}$
Zn *n*-amyl xanthate $(C_5H_{11}OCSS)_2Zn$	$S = 0.73 \times 10^{-4}$

Floatability obtained when benzyl xanthate is used as collector is analogous to that of butyl xanthate, However, the conjugated system is larger in $C_6H_4-CH{=}CH-CH_2OCSSNa$ than in C_6H_5. Comparison of thiophenols with various structures is listed in Table 5.14.

5.2.5.6. *Other non-polar groups*

Other non-polar groups in collectors include naphthalene and alkaryl groups. The naphthenic group exists in naphthenic acids and some xanthates. The length of a six member

Table 5.14
Properties of various thiophenols

Reagents	Structure	pK_a	Order of floatability
Butyl mercaptan	C_4H_9–SH	10.7	↓ Decreasing
Ethyl mercaptan	C_2H_5–SH	9.4	↓
Methyl thiophenol	$C_6H_5C_2$–SH	6.82	↓
Thiophenol	C_6H_5–SH	6.5	↓

naphthenic ring is 2.8 Å, which is analogous to ethyl or benzyl group. Their floatabilities are also similar.

Alkoxyl group exists in the non-polar position of some xanthates and dithiophosphates, which have properties similar to those of alkyl group and produce more froth. For example, etheralkyl carboxylate (ECA), R_{n_1}–$(OC_2H_4)_{n_2}$–OCH_2COOH, has been reported as a collector where n_1 is C_{8-18} chain and n_2 is the number of alkoxyl groups (0–16). Since its Ca or Mg salts with $n_1 = 10, 12, 14, 17$ and $n_2 = 10, 20\ 30$ are soluble in water, ECA is suitable under hard water and low pH (<6) conditions for the flotation of calcium or magnesium minerals, chalcopyrite and beryl with better selectivity. Floatability of Ca minerals is found to linearly decrease with n_1 values and increase with n_2 values.

5.3. FROTHERS AND MODIFIERS

5.3.1. Criteria of frother performance and interactions between frothers and other flotation agents

5.3.1.1. *Criteria for frother performance*

Frother performance can be evaluated by considering the following factors:

(a) Flotation (recovery and quality of the concentrate).
(b) Foam volume and life.
(c) Aeration rate of pulp and energy consumption.
(d) Size distribution of bubbles.
(e) Moving characteristics of bubbles (rising rate).
(f) Rate of coalescence of bubbles.
(g) Mechanical property of foam (tenacity, viscosity).
(h) Electrostatic properties of bubbles (zeta-potential).
(i) Interaction of frother with the mineral surface (coadsorption with collector).

It has been reported that in the case of flotation of galena, flotation performance depends on the micro-bubbles contained in the pulp as follows [17] (Table 5.14a).

Listed in Tables 5.15 and 5.16 are various properties of bubbles with frothers [7]. Dudenkov [2] classified frothers as two types with given fractions of 0.2 mm and −0.2 mm bubbles (Table 5.14b).

Table 5.14a

Particle size	Without micro-bubbles	With micro-bubbles
−0.15 + 0.044	21	31
−0.044 + 0.063	14	24

Table 5.14b

Type of frothers	Surface area (%)	
	−0.2 mm bubbles	+0.2 mm bubbles
Strong	70–90	20–50
Weak	15–50	80–90

Table 5.15
Size distribution of bubbles with various frothers (at 10^{-4} mol/l)

Frother	Size distribution (m.m. %)					
	+0.6	−0.6 +0.42	−0.42 +0.35	−0.35 +0.28	−0.28 +0.20	−0.20
Distilled water	22	26	25	16	9	2
Phenol	21	25	25	14	11	4
Methyl phenol	19	23	19	18	13	80
Dimethyl phenol	20	25	19	18	13	5
Aliphatic alcohol C_3	15	25	25	14	8	13
C_4	14	24	25	11	8	18
C_5	12	23	25	11	6	23
C_6	10	20	20	10	6	34
C_7	6	18	19	10	8	39
C_8	5	16	19	11	9	40
C_9	4	15	18	13	10	40
C_{10}	4	13	16	15	11	40
Cyclohexanol	13	25	20	17	9	18
Methyl cyclohexanol	12	22	13	12	6	35
Ethyl cyclohexanol	6	21	10	9	88	46
Propyl cyclohexanol	5	19	9	9	10	48
Polyglycol $-(O-C_4H_8)_n-$	–	3	3	5	12	77
$-(O-C_3H_6)_4-$	2	8	9	11	20	50
Alkoxyl triethoxybutane	–	5	5	7	15	70
$(CH_3)_2C_6H_2(COOH)_2$	10	21	10	18	12	23

In flotation, foam life and size distribution of the bubbles are mainly determined by the adsorption characteristics of the frother at the air/water interface including hydrophobic association between the chains, surface migration behavior of the molecule adsorbed in

Table 5.16
Dispersion state of bubbles with various frother (at 10^{-4} mol/l) (after Dudenkov [2])

Frother	Volume fraction of 0.2 mm bubbles	Surface area fraction 0.2 mm bubbles
Polyglycol $R–(O–C_4H_8)_2–OH$	0.23	0.830
Triethoxybutane	0.30	0.836
Polyglycol $R–(O–CH_2)_n–OH$	0.48	0.713
$R–(O–C_2H_5)_4–OH$	0.50	0.703
Aliphatic alcohol C_2	0.61	0.687
C_8	0.60	0.687
$C_4–C_8$	0.62	0.686
C_7	0.66	0.680
C_5	0.77	0.554
Dimethyl benzene dicarboxylic acid $(CH_3)_2C_6H_2(COOH)_2$	0.77	0.498
Cyclohexanol		0.457
Pine oil	0.83	0.478
Resol	0.92	0.250
Phenol	0.92	0.150
Distilled water	0.98	0.081

the liquid film and the zeta-potential of the bubble surface. These properties are closely dependent on surface viscosity of dilute solution and the adsorption density of frother.

5.3.1.2. *Effect of frothers on other flotation agents*

Possible interactions of frothers on the collector and other dissolved species include:

- Coadsorption of frother with collector on the mineral surface.
- Change in the surface tension of the flotation pulp.
- Enhancement of the collector role in frothing.
- Influence of inorganic ions on frothing.

The effect of surface tension change on flotation is shown in Fig. 5.9a and the effects of collectors and inorganic salts on frothing properties are illustrated in Figs. 5.9b and 5.9c.

5.3.2. Polar groups of frothers

The polar groups depending on its structure can be expected to influence the physical properties of frothers such as solubility, degree of dissociation and viscosity, and chemical properties such as bonding with minerals and reactions with the ions in the pulp. Characteristics of common polar groups of frothers are listed in Table 5.17.

5.3.2.1. *Polar groups and solubility*

Since the surface activity of highly soluble frothers is not very high, they require more dosages for sufficient frothing or multi-stage addition for keeping the foam alive. Frothers with low solubility need strong agitation for proper dispersion in the pulp. For longer life of

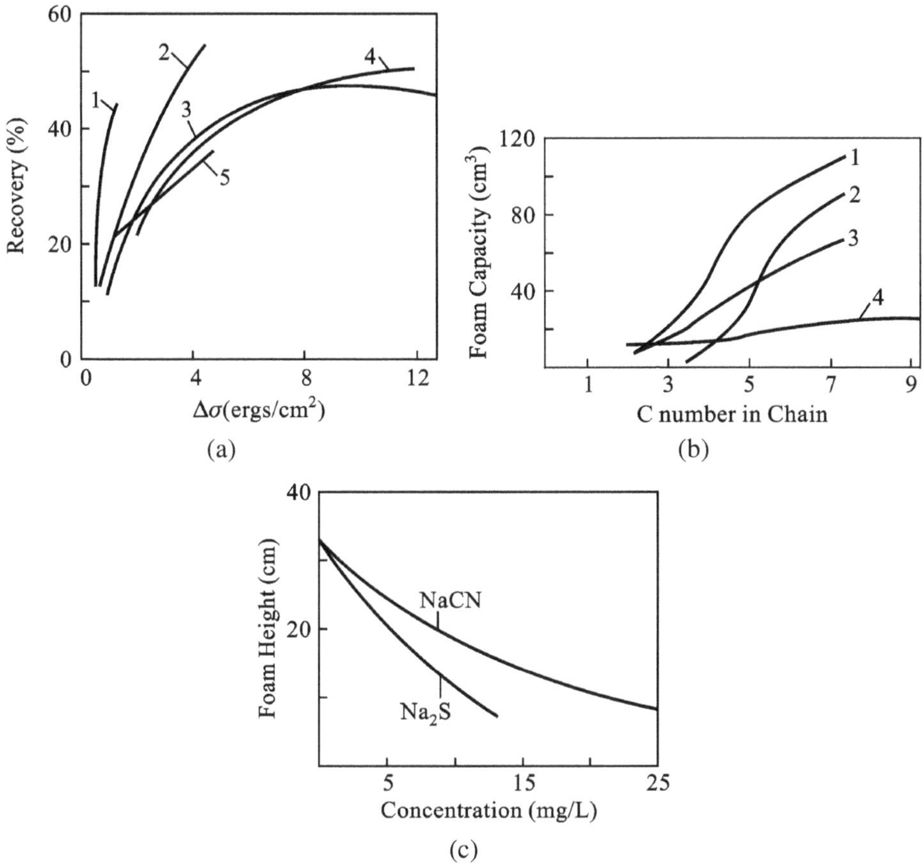

Fig. 5.9. (a) Influence of surface tension change on flotation of sphalerite and pyrite: 1—ZnS, FeS_2, quinoline; 2—ZnS, methyl quinoline; 3—ZnS, terpenol, cresol; 4—FeS_2, terpenol, cresol; 5—FeS_2, methyl quinoline. (b) Influence of xanthates on frothing of alcohols: 1—0.001 mol/l amyl xanthate + alcohols; 2—0.001 mol/l ethyl xanthate + alcohols; 3—only alcohols; 4—only xanthate. (c) Influence of inorganic salts on frothing of pine oil. Pine oil: 11.3 mg/l; $CuSO_4 \cdot 5H_2O$: 20 mg/l.

the foam, the chain length of less soluble frothers is limited to a narrow range. In addition, the required chain length of molecules of highly soluble frothers is longer, but the foam life may be short as shown in Fig. 5.10.

Poorly soluble frothers with long chains may have thicker adsorbed films and strong interactions between chains. Therefore, the stability and tenacity of foam can be expected to be high, however too long a chain can produce unstable films.

In flotation practice, less soluble frothers may also produce low frothing rate, finer bubbles and higher viscosity, and this can lead to higher flotation recovery. In contrast, soluble frothers exhibit fast frothing, coarser bubbles and brittle foam, and yield higher concentrate grade.

Table 5.17
Characteristics of polar groups of frothers

Polar groups	Interactions with		Foam stability & solubility
	Water molecules	Similar molecules	
$-CH_2I$, $-CH_2Br$, $-CH_2Cl$	Strong	Strong	No froth
$-CH_2OCH_3$, $-C_6H_5OCH_3$, $-COOCH_3$, $-OC_2H_3$	Strong	Weak	Not stable Low solubility
$-CH_2OH$, $-COOH$, $-CN$, $-CONH_2$, $-CHNOH$, $-C_6H_4OH$, $-CH_2COOH$, $-NHCONH_2$, $-NHCOCH_3$	Very strong	Intermediate	Stable & intermediate solubility
$-C_6H_4SO_3H$, $-SO_3H$, $-SO_4H$	Very strong	Intermediate	Stable & high solubility

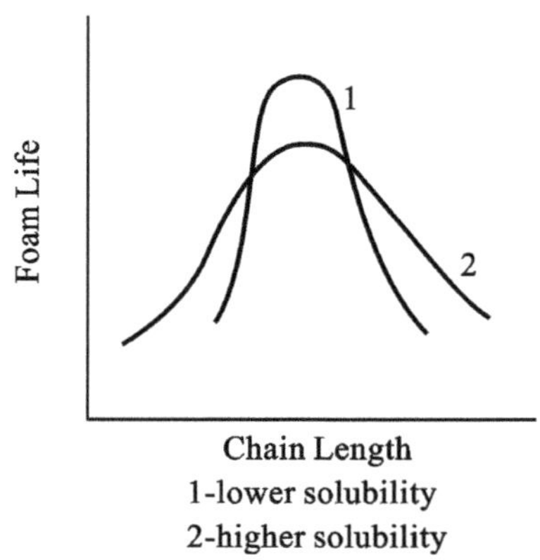

Fig. 5.10. The solubility of frother and their performance.

Solubilities of commonly used frothers are listed in Table 5.18. Solubility of frothers used commercially is about 0.0013%. Table 5.19 lists solubilities of frothers with given polar groups and various non-polar groups. It is shown that frothers with hydroxyl groups possess higher solubility.

5.3.2.2. *Degree of dissociation and frothing performance*

The pH value of medium controls the degree of dissociation of the ionic frothers and their dissolution. As a result, the frothing ability of ionic frothers is influenced markedly by the pH value of the media. Dissolution and the degree of dissociation of acidic frothers are high in alkaline media resulting in a decrease in the surface activity and the frothing ability. On the other hand, frothing ability of alkaline frothers increases in the alkaline media. This effect is illustrated in Figs. 5.11 and 5.12.

Table 5.18
Solubility of frothers (g/l)

Frother	Solubility	Frother	Solubility
n-amyl alcohol	21.9	*n*-TEB	0.586
i-amyl alcohol	26.9	*i*-TEB	1.28
λ-hexyl alcohol	6.24	Pine oil	2.50
MIBC	17.0	α-terpenol	1.88
MIBC	1.81	Camphor oil	0.74
i-MIBC	4.00	Cresol	1.66
TEB	~8	Polyglycol M.W. 400–450	Easily soluble

Table 5.19
Solubility of frothers with various polar and non-polar groups (mg/l)

Non-polar group	Polar group			
	–OH	$-NH_2$	–COOH	$-NO_2$
$-C_4H_9$	1055	–	333	–
$-C_6H_5$	874	383	24	15
$-C_6H_4CH_3$	202	–	7	5
$-C_6H_4CH_3$	227	158	9	–

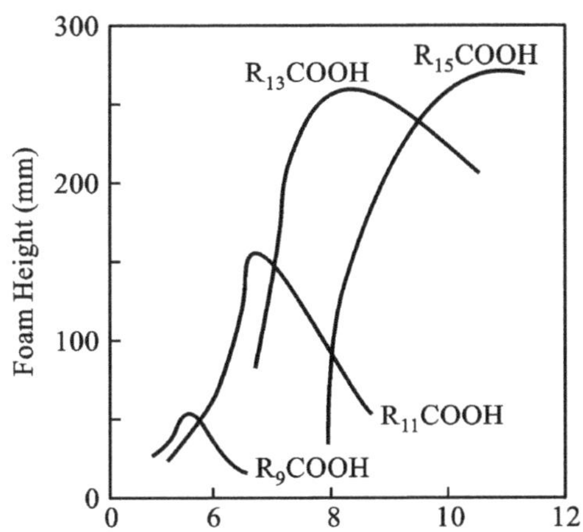

Fig. 5.11. Frothing power of fatty acid as a function of pH (concentration 0.1%).

As listed in Table 5.20, the electrostatic charge on bubbles is controlled by charge of the frother ions and the concentration of electrolyte in solution. In addition, the hydration of the frother molecule is important for the stability of foam bubbles. Interaction between polar groups of the frothers and water molecules can increase the strength of the film between bubbles.

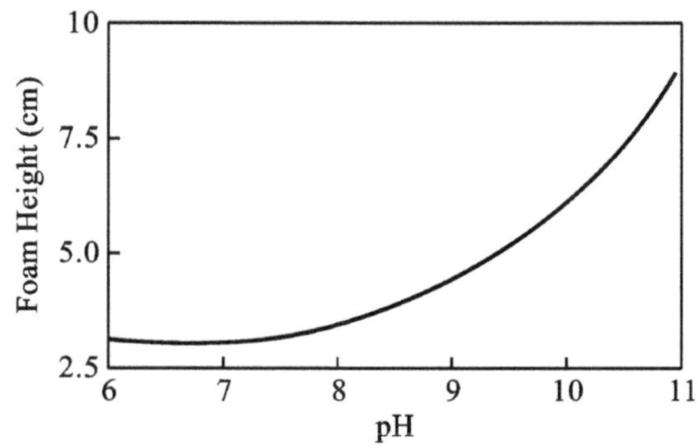

Fig. 5.12. Frothing power of pine oil as a function of pH values. Concentration: 11 mg/l (by Mitrofanov).

Table 5.20
Surface charge of bubbles in various electrolyte solutions

Electrolyte	Concn. (N)	Charge (V)	Electrolyte	Concn. (N)	Charge
KNO_3	0.01	−0.474	KI	0.001	−0.316
	0.0102	0.0		0.0075	0.0
	1.00	+0.068		1.00	+0.078
KCl	0.01	−0.294	$Al(NO_3)_3$	0.001	−0.024
	0.012	0.0		0.0203	0.0
	1.00	+0.079		1.022	+0.064
KBr	0.001	−0.344	$Mg(NO_3)_2$	0.0012	−0.562
	0.014	0.0		0.0025	0.0
	1.00	+0.080		1.2448	+0.058

5.3.3. Non-polar groups of frothers

5.3.3.1. *Normal alkyl group*

As stated above, the surface activity of a surfactant with normal alkyl chain in its molecule increases 3.2 times because of the addition of every –CH= group in the molecule. Table 5.15 shows that with the increase in the chain length especially in the range of carbon number of C_3–C_8, fraction of the fine bubbles increases.

5.3.3.2. *Iso-alkyl and unsaturated chain*

Frothers with highly branched chains possess high frothing ability. Such frothers which are widely used in flotation are methyl isobutyl carbinol and triethoxybutane:

MIBC: $H_3C–CH(CH_3)–CH_3–CH(OH)–CH_3$

TEB: $CH_3–CH(OC_2H_5)–CH_2–CH(OC_2H_5)_2$

The floatability of terpinol and resinolic acid are improved with the introduction of double bonds in molecules. Conversely, if the existing double bonds are eliminated, the result would be a decrease in frothing ability. Natural frothers with double bonds, branched chain or unsymmetrical structures such as terpinols, resinolic acids usually possess good frothing ability.

5.3.3.3. *Aromatic and alkyl-aromatic groups*

Commonly used aromatic frothers are phenols, cresols, pyridines. These are mixtures of various methylphenols, naphthols, quinolines, anilines, naphthalene, amines, etc. extracted from coal tars. These frothers have been widely used in the flotation of coal. Apart from the frothability, these reagents have certain ability as collector especially with the increase in their molecular size.

Commonly used alkyl-aromatic frothers are alkyl-benzene sulfonates (sodium salt) which possess not only frothability but can also be used as collectors especially for sulfide minerals. The effect of an alkyl chain linked with benzene ring on frothing performance is an increase in the frothability with increase in the length of alkyl chains of C_5–C_{12} over which the frothability will usually decrease.

5.3.3.4. *Alkoxyl group*

General formula is as follows:

$R–(OC_nH_{2n})_m–$

By introducing $–CH_2–$ groups among with an –O– group, the following characteristics are obtained.

In the presence of an –O– group, the polarity, hydrophilicity and solubility in water is higher than those of the same $–CH_2–$ number of alkyl groups. The surface activity of the reagents increases with the increase of n in a certain range.

Hydrophilization of –O– group in the non-polar portion of the molecule is brought about by the interaction with water molecules through hydrogen bonding and van der Waals force. Such hydrophilization decreases with increase in temperature. In aqueous solution, the configuration of the chain is tortuous with the hydrophilic oxygen atoms outside and the $–CH_2–$ groups inside, resulting in hydration of the chains. As the temperature is raised, the water molecules attached to the chains are removed and the frother molecules become less soluble in water. In addition, interaction of –O– group with other reagents becomes stronger leading to higher coadsorption.

5.4. DEPRESSANTS

5.4.1. Classification

Depressants are usually divided into three types:

(a) Inorganic depressants: acids, bases and salts.
(b) Low molecular weight organic depressants: containing –OH, –COOH, –SH or =S, $–NH_2$ and $–SO_3H$ groups.
(c) High molecular weight organic depressants: reagents such as tannin, starch, cellulose, lignin, gum and polyacrylic acids.

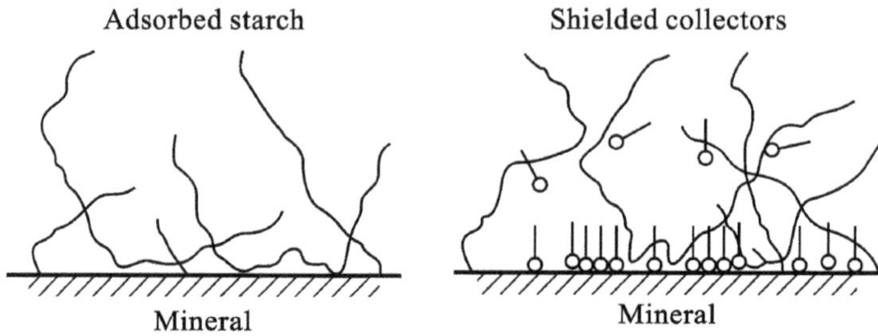

Fig. 5.13. Depression of polymer (by Somasundaran).

5.4.2. Depression mechanisms

Use of various reagents for the depression of mineral flotation is often needed for the following purposes:

(a) pH and E_h control in flotation pulp [26]: pH value in the pulp directly determines the hydration, oxidation–reduction and the double layer structure of mineral surface, thereby regulating the hydrophilic properties of mineral surfaces and their interaction with collectors or other reagents. Another important role of pulp pH is to control chemical state of the species in aqueous solution resulting from mineral dissolution and dissociation, hydrophobic association with polymers or micelles and oxidation–reduction reactions. Recent studies have shown that E_h control by oxidizing or reducing agents are very important, especially for the flotation of sulfide minerals and for collectorless flotation.

(b) Regulation of ionic components in the pulp: dissolved mineral species such as Cu^{2+}, Fe^{2+}, Pb^{2+}, Ca^{2+} and Mg^{2+} can change flotation behavior of minerals, through activation, depression or by precipitation of the collectors, all of which can result in poor separation.

(c) Hydrophilization of mineral surfaces by direct adsorption of depressants: inorganics such as sodium sulfide, sodium silicate and sodium cyanide and organics such as hydroxyl acids and thioglycols are often used as depressants.

    Polymeric depressants adsorbed on mineral surfaces can directly cause the hydrophilization since their long chains can mask the adsorbed collectors. Somasundaran et al. [14] have investigated depression of calcite by starch. Their work has revealed that the starch does not decrease the adsorption of oleic acid on the mineral, flotation is prevented by hydrophilization by starch being more than hydrophobization by oleic acid as shown in Fig. 5.13.

(d) Desorption of collector from the mineral surface or collector coadsorption: competitive adsorption of similarly charged ions is an important mechanism of depression. It has been reported that depression of feldspar by sodium salicylate is due to the retardation of the adsorption of oleic acid on the mineral surface [4]. In the flotation system of hematite–oleic acid–tannin, it is observed that while oleic acid or tannin reacts individually with hematite with significant flotation effects, their adsorption is reduced

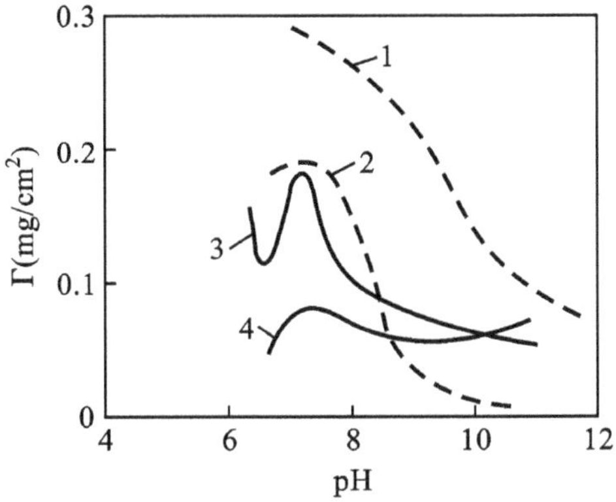

Fig. 5.14. Adsorption of oleic acid and tannin on hematite: 1—tannin S 40 mg/l; 2—tannin 40 mg/l (oleate 5×10^{-5} mol/l); 3—oleate 5×10^{-5} mol/l; 4—tannin S 40 mg/l (oleate 5×10^{-5} mol/l).

when both the reagents are present. This indicates clearly the role of the competitive adsorption between the collector and the depressant (see Fig. 5.14).

Various interactions of mineralophilic groups of depressants on the minerals can be summarized as follows:

(a) Adsorption in the double layer on the mineral surface: one of the examples in this category is modified ionic starches which are able to adsorb on negatively charged quartz and cationic starches to adsorb on positively charged hematite.
(b) Adsorption through hydrogen bonding and van der Waals forces: many hydroxyl, carboxyl and sulfonate organic depressants are known to adsorb on the mineral surfaces as a result of hydrogen bonding. For example, starch derivatives, especially natural starch and unmodified dextrin products have been considered to adsorb through hydrogen bonds between hydrogen atoms in the hydroxyl groups of the starch molecules and the oxygen atoms of minerals as in the case of minerals such as quartz. The magnitude of free energy of adsorption of starch on minerals, about 5 kcal/mole, is a characteristic of hydrogen bonding and van der Waals force.
(c) Chemisorption and other surface reactions: certain organic depressants possess chemically active groups such as thiol, carboxyl and amino groups, which are capable of reacting with elements on the mineral. Hence, chemisorption or such chemical reactions on the surface can be major mechanisms for adsorption of these reagents. Investigation by Schubert on adsorption of collectors and depressants on fluorite has shown that adsorption of various reagents is in the order [13]:

$$C_{12}H_{25}NH_3^+ < C_{12}H_{25}OSO_3^- < C_{11}H_{23}(COO^-)_2 < C_{10}H_{21}CH(COO^-)_2$$
$$< \text{alizarin} < \text{citric acid} < \text{Congo Red}$$

Chemisorption of dicarboxylic acid as a collector is considered to be strong, but these results show that the above depressants can be even more strongly adsorbed on the minerals. The above mechanisms are considered to be part of the adsorption of many other

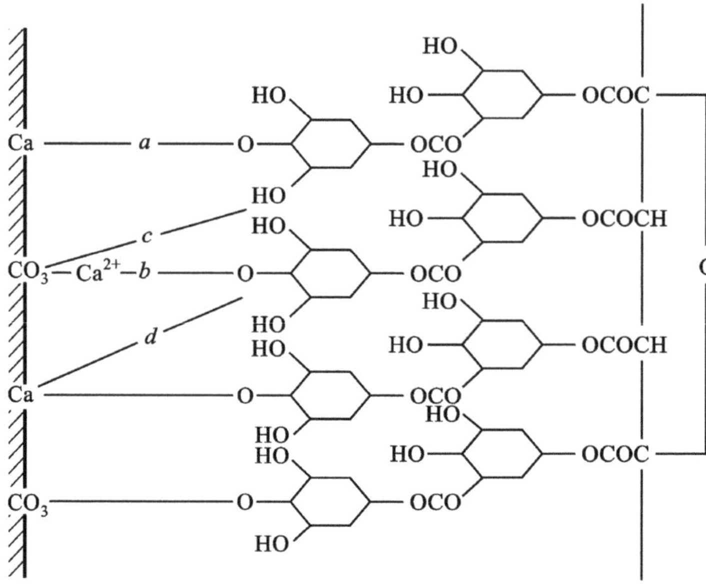

Fig. 5.15. Adsorption of tannin as a depressant on calcite. (*a*) Interaction between –OH group of tannin and Ca ions on mineral. (*b*) Adsorption of the depressant on the Ca ions on the mineral depressant. (*c*) Hydrogen bonding. (*d*) Reaction due to electrostatic forces.

reagents. Somasundaran et al. [25] proposed the adsorption of tannin as a depressant on calcite to be due to (Fig. 5.15).

5.4.3. Inorganic depressants

Inorganic depressants are usually used in combination with other flotation agents. Inorganic reagents used as depressants are listed in Table 5.21 and commonly used ones are classified below according to their application:

(a) For depression of copper minerals: P–S compounds, for example, called "Notes Depressants", sulfides, oxidizing agents such as air, $KMnO_4$ and H_2O_2 have been used for primary copper minerals, and cyanides like $K_4Fe(CN)_6$ and $Zn(CN)_2NH_3$ for secondary copper minerals.
(b) For depression of PbS: reducing agents especially S–O salts such as H_2SO_3, SO_2, $Na_2S_2O_3$, $K_2Cr_2O_3$, sulfides, iron salts and phosphates.
(c) For depression of ZnS: compounds such as H_2SO_3, SO_2, Na_2SO_3, $ZnSO_4$, CaO, sulfides and cyanides.

5.4.4. Structure of organic depressant

The structural characteristics of organic depressants may be summarized as follows: the molecule must have as a small hydrocarbon framework as possible with several (over two) polar groups; the polar groups located anywhere at the two ends; a fraction of the

Table 5.21
Inorganic reagents commonly used as depressants

Type		Depressants	Application
Anionic	Cyanide	$NaCN$	For depression ZnS, FeS_2
		$Ca(CN)_2$	Idem.
	Ferricyanide	$K_4Fe(CN)_6$	Idem.
		$K_3Fe(CN)_6$	Idem.
	Zinc cyanide	$K_2Zn(CN)_4$	Idem.
		$Zn(CN)_2(NH_3)_2$	Idem.
	Sulfide	Na_2S	For depression ZnS, FeS_2
		$NaHS$	Idem.
		Na_2S_x	Idem.
		CaS_x	Idem.
	Acid containing S	H_2SO_3	ZnS
		Na_2SO_3	Idem.
		SO_2	Idem.
		$Na_2S_2O_3$	Idem.
		HOC_2SNa	Idem.
		$(HO)_2PS_2Na$	Sulfide minerals
	Chromate	K_2CrO_7	Idem.
	Silicate	Na_2SiO_3	Silicate minerals
		Na_2SiF_6	Idem.
	Fluoride	HF	Silicate minerals
		NaF	Idem.
	Phosphate	Na_3PO_4	Gangue minerals
		Na_2HPO_4	Idem.
		$(NaPO_3)_n$	Idem.
Cationic	Zinc salt	$ZnSO_4$	ZnS
	Calcium compound	CaO	FeS_2
		$CaCl_2$	Idem.
		$CaOCl_2$	CuS and iron minerals
	Iron salt	$FeSO_4$	Idem.
		$Fe_2(SO_4)_3$	Idem.
	Aluminium salt	$Al_2(SO_4)_3$	Idem.

polar groups being minerophilic and other groups projecting outward so as to bring about hydrophilization of the mineral. The bonding of the minerophilic group must be stronger than that of the collector and the hydrophilic groups must have high hydration power.

The minerophilic groups of various collectors may also be used as depressants as in, for example:

- thiocarbonates: HOCSSH, HSCSSH;
- thiophosphates: $(HO)_2PSSH$, $(HO)_2POSH$;
- thiocarbamates: NH_2CSSH, $(NH)_2PSSH$;
- thio groups: –SH, $(NH_2)_2CS$, $NH_2C(SH)_2$
- carboxylates: –COOH, –CH(OH)COOH, $–CH(NH_2)COOH$
- sulfonates, sulfates, phosphates, etc.

The hydrophilic groups commonly used are –COOH, –OH, –O– and $–SO_3H$.

The structure–property relationship of the depressants used in some flotation systems are discussed below.

5.4.4.1. *Structure of organic depressants for Cu^{2+}-activated quartz [27]*

(I) Benzoic acid

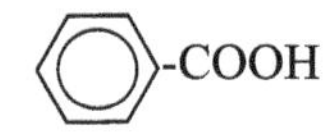

(II) Phthalic acid

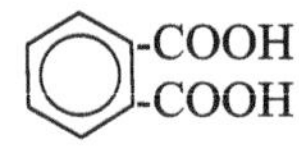

(III) Salicylic acid

(IV) Methyl salicylate

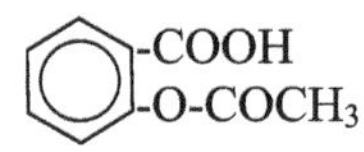

(V) Phenyl salicylate

The order of the ability of the above reagents to depress (depressability) is:

(II) > (III) > (I) > (IV) > (V).

Reagent (V) does not seem to have any role in causing depression and reagent (IV) depresses the flotation only at high dosages. It must be noted that the depressability of the reagents increases with increase in the number and activity of the polar groups. Hydration and adsorption capacity of these groups have been discussed in Sections 5.1 and 5.2. The order of the depression obtained with the above three groups is –COOH > –OH > $–COOCH_3$. Reagent (II) with two –COOH radicals is the most powerful agent while reagent (III) with one –COOH and –OH groups is the next. Reagents (IV) and (V) have only a –OH group yielding low depression.

The order of depression of these reagents when used on Ca^{2+}-activated quartz is as follows: (XI) > (X) > (VII) > (VIII) > (IX) > (VI).

Citric acid with three –COOH and one –OH groups and tartaric acid with two –COOH and two –OH groups are excellent depressants, oxalic and amberic acid possessing two COOH groups also exhibit good depression, but the two $–CH_2–$ groups on the latter molecule reduces its hydrophilicity; acetic acid and lactic acid which have only one –COOH group are very weak depressants.

Comparison between aromatic and aliphatic acids: dosages (mol/t) of the above depressants required for depression of 96–99% quartz are used as the criterion for evaluating the reagent power. The order for the above reagents are as follows (the data in brackets are dosages):

$$(XI)_{(2.5)} > (X)_{(4.0)} > (II)_{(4.5)} > (III)_{(9.0)} > (VII)_{(11.0)} > (I)_{(12.3)} > (VIII)_{(12.75)} > (IX)_{(17.6)} > (VI)_{(33.2)}$$

This order shows that the depressability of the reagents with aromatic framework is higher than that of the reagents with aliphatic framework having the same polar groups.

5.4.4.2. *Organic depressants for fluorite flotation [18]*

(I) Citric acid
(II) EDTA-Na
(III) Tartaric acid
(IV) Maleic acid
(V) Butyl dicarboxylic acid
(VI) Lactic acid
(VII) Benzene dicarboxylic acid
(VIII) Congo Red
(IX) Alizarin Red
(X) Gallic acid

When a mixture of sulfonate and sulfate is used as the collector, the order of depressability of the above reagents is: (I) > (II) > (IV) > (III) > (X) > (VII) > (VI) > (V). This order shows that higher depression ability is possessed by those molecules which have more polar groups, higher activities of such groups and less non-polar portion.

Adsorbability of various reagents and collectors on fluorite is as follows: (III) < (VI) < (V) < (VII) < (IX) < (I) < (VIII).

Depressants can impede adsorption of collectors on the mineral surfaces in the following order: citric acid > Alizarin Red S > EDTA-Na. But adsorption of these depressants themselves on the minerals is in the order of: (VIII) > (I) > (VII) > (III) > (VI). These order show that adsorbability of effective depressants is higher than that of carboxylate collectors which chemisorb on mineral surfaces.

5.4.4.3. *Organic depressants for Fe^{3+}-activated wolframite and quartz*

Depression of wolframite–quartz flotation with various flotation agents is shown in Figs. 5.16a and 5.16b. Experimental results for the concentration of the reagents for good depression of the above two minerals are listed in Table 5.22. The structural characteristics of the depressant molecules can be summarized as follows:

(a) The depressability of organic depressants is determined by the number of polar groups in their molecules. Reagents with only one polar group do not possess any ability to depress. Those that exhibit depression property will have two or more polar groups some of which are minerophilic and others are hydrophilic.
(b) Depressability also depends on the reactivity of the polar groups towards water or minerals.
(c) furthermore depressability is influenced by the extent to which the polar groups in a molecule project outwards or inwards. This is exemplified by the fact that amine groups in EDTA molecules do not play any role in depression.

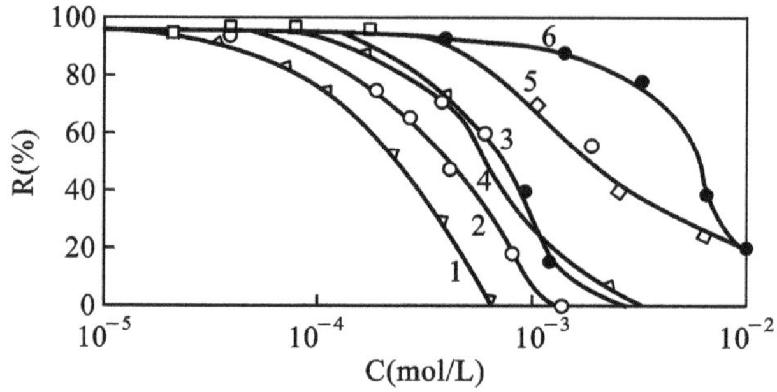

(a)

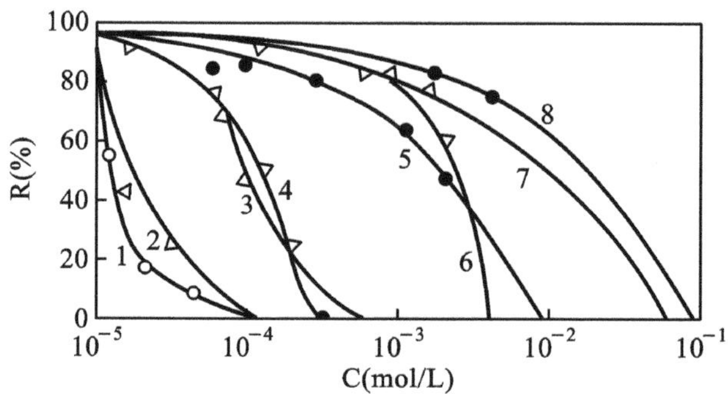

(b)

Fig. 5.16. (a) Depression of various reagents on wolframate flotation: 1—citric acid, 2—tartaric acid, 3—oxalic acid, 4—EDTA, 5—succinic acid, 6—lactic acid. (b) Depression of various reagents on flotation of Fe^{3+} activated quartz: 1—tartaric acid, 2—citric acid, 3—EDTA, 4—oxalic acid, 5—succinic acid, 6—butyl dicarboxylic acid, 7—lactic acid, 8—acetic acid.

Table 5.22
Concentration of reagents for depression of minerals, mole/l

Reagents	For quartz	For wolframite
Citric acid	4.0×10^{-5}	4.6×10^{-4}
Tartaric acid	3.0×10^{-5}	2.0×10^{-4}
EDTA	1.3×10^{-4}	1.2×10^{-4}
Oxalic acid	2.2×10^{-4}	1.0×10^{-3}
Butyl dicarboxylic acid	4.0×10^{-3}	–
Succinic acid	4.0×10^{-3}	3.0×10^{-2}
Lactic acid	6.0×10^{-3}	3.0×10^{-2}
Acetic acid	1.5×10^{-1}	–

(d) Depression by the reagents with aromatic framework is always higher than those with aliphatic ones. In addition to the above, there is another class of organic depressants which are used as oxidizing or reducing agents to control E_h potential of the pulp or the redox reaction of the collector on the surface of sulfide minerals.

The hydrophilicity of the polar groups of the above reagents is not very high and the depression is caused mainly through the chemical activity or nature of the reagents.

5.4.5. Structure of polymeric depressants

5.4.5.1. *Structural characteristics of polymeric depressants*

Their characteristics listed in Table 5.23 include those of various types of polar groups and C–H frameworks. Two special mechanisms of depression of polymeric depressants are described below:

(a) As polymeric depressants possess long and branched chains, they can get adsorbed on mineral surfaces not only maintaining their hydrophilicity but also masking collectors that are adsorbed on such surfaces and thus restoring their hydrophilicity. They can also indeed prevent adsorption of the collectors.
(b) Polymeric depressants have both depression and flocculation properties with the result that the particles are aggregated and the flotation behavior of the minerals is changed.

5.4.5.2. *Structure–property relationship of natural starch [17]*

Various starches have been used in industry for depressing talc, mica, natural sulfur, carbon gangue and sulfide minerals, and especially for depressing oxidized iron minerals in the reverse flotation of iron ore and the separation of Cu–Mo in the copper–molybdenite sulfide flotation.

Starches made from stem tubers of plants have various chemical components and structures. According to the different configurations of chains in their molecules, starch can be divided into amylose (with straight chain) and amylopectin polymer (branched or gelatinized). The characteristics of these two kinds of starch are as follows (Table 5.23a).

Common natural starch contains less of amylose than the amylopectin one, and 3–25% of the former is wrapped within a framework of the latter. Composition of starches from different plants with respect to the above two classes are shown in Table 5.23b.

5.4.5.3. *Modified starch*

A common example of the modified starch is dextrin made by the hydrolysis of natural starch with dilute acids. Its molecular weight is lower than that of the natural starch. In the literature, the dextrin is also called British gum which has a molecular weight of about 800–79 000. Application of dextrin is similar to that of starch as a depressant for molybdenite and hematite used alone or in combination with other reagents such as aniline employed for depression of gangue slime and carbon gangue. It has been reported that adsorption of dextrin on molybdenite conform to the Langmuir equation with the free energy of adsorption as 5.4 kcal/mol-dextrin unit. Since dextrin is nonionic, its adsorption on the minerals has no direct influence on the double layer of the mineral surface. While the relative value of the zeta-potential does change with the addition of dextrin, there is no change

Table 5.23
Structural characteristics of commonly used polymeric depressants

Reagent	Structural characteristics		
	Polar group		Framework
Starch			
Natural	–O–, –OH	α-glucose	Corn, cassava for hematite & MoS_2
Cationic	$-N(CH_3)_2$, –O–	...	Silicate
Anionic	–COOH, –O–	...	Hematite & gangues
Dextrin	–O–, –OH	...	...
Cellulose			
Carboxyl methyl	–COOH(Na), –O–	β-glucose	Ca–Mg minerals & slimes
Hydroxyl ethyl	–OH, –O–	...	...
Sulfonate	$-SO_3HNa$, –O–	...	...
Xanthate	–OCSSHNa, –O–	...	Sulfides
Polyglucose			
Natural gums	–O–, –OH	Glucose	Ca–Mg minerals & slimes
Modified	–O–	...	...
Lignin			
Natural	–OH	$-C_2H_4-$, $-C_3H_6-$	RE minerals & gangues
Sulfonate	–SOH(Na,Ca), –OH	...	...
Chloride	–Cl, –OH	...	...
Tannin			
Natural	–COOH, –OH	$-C_6H_4-$, etc.	...
Oxide	...	...	...
Sulfide	... & –SH, =S	...	...
Synthetic	$-SO_3H$, –Cl	...	...
Acrylic polymers			
Polyacrylic acid	–COOH	$-CH_2-$	Flocculant for gangues
Polyacrylamide			

Table 5.23a

	Amylose	Amylopectin
Linking position	1,4 chain	1,6 or 1,3 chain
Polymerization degree	200–980	600–6000
Molecular weight	32 000–160 000	10^5–10^6
Solubility in water	Soluble	Limited

Table 5.23b

Plants	Amylose	Amylopectin
Corn	15	85
Potato	3	97
Rice	12	88
Wheat	23.5	76.5

with the solution pH itself. It has been suggested that adsorption of dextrin on the minerals is mainly due to hydrogen bonding and van der Waals forces.

Modification of starch molecules is carried out by the addition of various chemical radicals through oxidation, etherification, esterification, and introduction of anionic or cationic groups, generally at 2 and 6 sites. Major chemical reactions involved in the preparation of modified starches are given below.

Phosphate starch (anionic):

$$\text{Starch–OH} + Na_5P_3O_{10} \rightarrow \text{Starch–OPO}_3\text{Na} + Na_3HP_2O_7 \qquad (5.8)$$

Alkoxyl starch (nonionic):

$$\text{Starch–OH} + \underset{\diagdown\;\diagup \atop \text{O}}{H_2C\text{—}CH_2} \rightarrow \text{Starch–OCH}_2\text{CH}_2\text{OH} + n\underset{\diagdown\;\diagup \atop \text{O}}{H_2C\text{—}CH_2}$$
$$\rightarrow \text{Starch–O–}(CH_2CH_2O)_n^{-}CH_2CH_2OH \qquad (5.9)$$

Tertiary amine starch (cationic):

$$\text{Starch–OH} + ClCH_2CH_2N(CH_3)_2 + NaOH \rightarrow \text{Starch–OCH}_2\text{CH}_2\text{N(CH}_3)_2 + NaCl + H_2O \qquad (5.10)$$

Depression mechanisms of various modified starches are different from that of each other. In aqueous solution, starch molecules are associated with each other by means of hydrogen bonding. While the unmodified starch is negatively charged in the pH range of 3 to 11, the charge of modified starch molecules depends on the substituted groups in them.

As expected, cationic starches adsorb on negatively charged mineral such as quartz while anionic starches adsorb more on positively charged minerals such as hematite in the low pH range. Since mineral surface will become more negatively charged with increase in pH, the adsorption of anionic starches decreases with rising pH and that of cationic starches increases. Natural starches are slightly negatively charged and hence adsorb more on hematite than on quartz in the low pH range and more in the lower pH range than in the higher pH range. On the contrary, adsorption of the cationic starch on hematite is lower than that on quartz and increases with increase in pH.

In summary, adsorption of starches on minerals takes place through hydrogen bonding, the van der Waals force and electrostatic force while chemisorption also plays a role in the area of adsorption of modified starches.

5.4.5.4. *Tannin*

Tannin, which is available as dry powder or as concentrated aqueous solution, is extracted from plants commonly called quebracho. Often the portions of the plant containing large amounts of tannin are broken up directly to make tannin powders. While the composition of the various quebrachos can vary greatly depending on their source and processing, basic structural units are all phenol groups linked to the molecule that have a molecular weight of 2000 or more.

Tannins are usually divided into two kinds depending on the components and properties: soluble gallic acid and cationic acid. They are obtained through heating and hydrolysis in dilute inorganic acid, and are called hydrolyzed tannin and agglomerated tannin, respectively.

Tannin is soluble in water and alcohol. It can form colored complexes or precipitates with calcium, iron and lead ions. The tannin that is commonly used as a depressant is called quebracho, which is extracted from apocynaceae, a plant species in South America. It contains agglomerated tannin.

Tannins are mainly used as depressants for dolomite, calcite and silicate minerals in the flotation of scheelite, apatite, fluorite and copper sulfide. Like starch, tannins after modification by oxidation, sulfidization and aminization (tannin O, S and A, respectively) have been used as a depressant for hematite in the reverse flotation of quartz with oleic acid or dodecylamine as collectors [22]. Flotation results show quartz to be depressed by tannins in the order of tannin A > S at low concentrations of dodecylamine. When an anionic collector is used for the reverse flotation of Ca^{2+}-activated quartz with tannins as depressant for hematite, the order of depression is tannin A > O > S.

5.4.5.5. *Lignin*

Lignin is one of the complex organic substances occurring as a major constituent in the stems, roots and leaves of trees and plants, with about 60–70% of wood being lignin. Essentially, lignin is a natural polymer containing a variety of isomers with –OH and $-OCH_3$ groups. This polymer possesses a lower degree of polymerization than cellulose along with which lignin invariably occurs in nature and has a molecular weight of about 800 to 1000.

Materials used for obtaining lignin depressants are usually wastes and by-products of various industrial processes which use wood and seeds as raw materials. Compounds such as soluble lignin sulfonic acid, chloride and alkaline lignin can be made by treating lignin with various chemicals. The polar groups in lignin may be added on to the branched chains of the propane or the phenyl ring.

Selective flotation of fluorite and barite is usually very difficult using water glass and tannic acid. However, good separation of these minerals has been achieved with the use of lignosulfonate [18]. Lignosulfonate can also selectively depress molybdenite and some rare earth metal minerals. In the reverse flotation of hematite and quartz, lignosulfonate has been used as a depressant of hematite.

5.4.5.6. *Gums*

Gums are macromolecular substances containing polysaccharides extracted from natural plants. Gums can be used as depressants directly or after chemical modification. Guar gum which is a depressant for silicate minerals is a product obtained from the processing of the guar seeds common to the Arabic countries and India.

Gum arabic is a mixture of Ca, Mg and K salts of Arabic saccharinic acid which consists of 1-D-hexanic acid, 3-D-galactose, 2-L-arabic saccharide and 1-L-rhamnose. The saccharosan is contained in seeds, bud stems, roots, leaves and skins of many plants. In general, all the gums, including starches and cellulose, are natural polymers of the saccharide units although their structures are quite different from one another. Starch has a simpler struc-

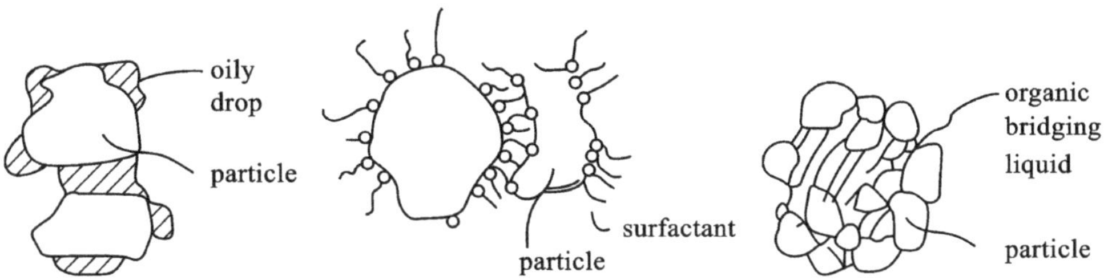

Fig. 5.17. Hydrophobic agglomeration process.

ture with less number of branches in the molecules while cellulose has a more twisted structure and gums have more complex and branched structure. They all are capable of conversion to macromolecules containing –COOH and –OH groups by appropriate chemical processes. Such modified gums find application in the flotation of various minerals. For example, etherized gum is mainly used as a depressant for silicate minerals such as talc, mica, serpentine and chlorite and also in the flotation of potash, Ni sulfide and other nonferrous metal sulfides.

5.5. FLOCCULANTS

5.5.1. Types of flocculants and flocculation

In general, flocculants are used in solid–liquid separation processes such as thickening and filtration. Inorganic salts are also used sometimes to aggregate fine particles. Flocculation technique has been developed further for special applications of selective flocculation, selective dispersion and agglomeration flotation.

Aggregation of fine particles can be classified mainly into four types:

(a) Coagulation by electrolytes: settling of fines can be achieved by the addition of electrolytes which can decrease the zeta-potential of the particles and the electrostatic repulsion between them and resultant aggregation of the particles in the medium. Inorganic compounds and polymeric electrolytes usually belong to this category.
(b) Hydrophobic agglomeration: mineral particles possessing hydrophobicity on the surface are agglomerated in this case owing to association of hydrocarbon chains of adsorbed surfactants and non-polar oil (in Fig. 5.17). If sufficient oil is added, hydrophobic particles are wetted by the oil which form an organic bridge between them. These agglomerates are not only large but also possess high hydrophobicity and floatability.
(c) Bridging by polymers: polymer molecules possess long chains which contain number of minerophilic groups so that they can get absorbed on the particles and lead to bridging them by polymer chains and consequently to flocculation.
(d) Aggregation of particles by magnetic force or other forces: in a magnetic field, fine particles of magnetic minerals such as magnetite are aggregated by the force produced

Table 5.24
Types of flocculants

Reagents	Mechanism	Structure	Examples
Soluble inorganic compound	Coagulation	Inorganic salts, acids, bases	$Al_2(SO_4)_3$, $Fe_2(SO_4)_3$, $FeSO_4$ $NaAlO_2$, $FeCl_3$, $ZnCl_2$, $TiCl_4$, H_2SO_4, HCl, CO_2, NaOH, CaO, Na_2CO_3
Inorganic colloid particles	Neutral charge of particles and coprecipitation	Hydrolysis products of metallic ions, solid powder, inorganic polymer	$Al(OH)_3$, $Fe(OH)_3$
Organic surfactants	Hydrophobic agglomeration	Heteropolar molecule	Sodium oleate, sulfonate, xanthate
Hydrocarbon oil	Oil bridging	Non-polar molecule	Kerosene, fuel oil
Organic low M.W. polymer	Bridging and coagulation	Anionic	Alginate-Na, carboxyl-methyl cellulose-Na, sodium humate
(M.W. thousand to tens of thousands)		Cationic	Soluble aniline resin, polythiourea, polyvinyl ethylamine
		(P225)	Poly-2-methacryloyloxy ethylene Trimethyl ammonium chlorate
		Amphoteric	Animal gelatins
		Nonionic	Starches, gums
Organic high M.W. polymer	Bridging and coagulation	Anionic	Polyacrylic acid (-Na) Hydrolyzed polyacrylamide
		Cationic	Polyethylene pyridine chloride
		Nonionic	Polyacrylamide, polyhydroxyethylene

by a magnetic equipment or magnetic particles present in the pulp. Commonly used flocculants are listed in Table 5.24.

Coagulation by inorganic electrolytes is caused by adsorption in the double layer of the mineral particles and resultant reduction in the magnitude of the zeta-potential into the range for instability, i.e. usually from 0 to 15 mV in magnitude. The coagulation power of these reagents depends on the properties of relevant counterions of the electrolyte as well as the mineral surface. Types of inorganic ions that can act as coagulants are discussed below.

5.5.2. Inorganic electrolytes as coagulants

(a) Potential-determining electrolytes: an electrolyte containing the potential-determining ions for the mineral particles can adsorb on the surface deep within the Stern plane of the double layer or as a part of the mineral crystal lattice and influence the magnitude of the zeta-potential drastically. Therefore, the potential-determining ions which are the

Table 5.25
Potential determining ions, point of zero charge (PZC) and solubility products, pK_{sp} of minerals

Minerals	Potential determining ions		Point of zero charge		pK_{sp}
$BaSO_4$	Ba^{2+},	SO_4^{2-}	pBa	6.7	1.5×10^{-9}
CaF	Ca^{2+},	F^-	pCa	3	1.7×10^{-10}
$CaWO_4$	Ca^{2+},	WO_4^{2-}	pCa	4.8	8×10^{-3}
AgCl	Ag^+,	Cl^-	pAg	4	1.56×10^{-10}
AgI	Ag^+,	I^-	pAg	5.6	1.5×10^{-16}
Ag_2S	Ag^+,	S^{2-}	pAg	10.2	1.6×10^{-49}

mineral lattice constituent or isomorphous substituents may be used as an effective coagulant. Listed in Table 5.25 are certain potential-determining ions and concentrations at the point of zero charge (PZC) of the minerals.

(b) Inorganic acids and bases: hydrogen and hydroxyl ions are potential-determining ions for certain minerals such as oxides, silicates and salts of oxygen-containing acids. Therefore, pH of the medium has a major influence on the surface charge density of the mineral particles, and can cause their coagulation at certain value.

(c) Indifferent electrolytes: ions adsorbed to maintain the electroneutrality of the mineral surface are called counterions and most inorganic ions charged oppositely to the mineral surface can function as counterions. Based on the Gouy–Chapman treatment of the double layer, the relationship between the concentration of the counterions, zeta-potential value and diffuse layer charge density is given as follows:

$$\sigma_d^+ = \sqrt{2\varepsilon nkT/\pi}\,\sinh(ze\zeta/2kT), \tag{5.11}$$

where σ_d^+ is charge density of the diffuse layer of a negatively charged mineral particle; n the number of ions per cm^3 in solution; z the valency of the ion; e the electronic charge and ε the dielectric constant.

It can be seen that the effect of indifferent electrolytes on the zeta-potential of the mineral particle depends not only on its concentration but also on the valency of the counterions involved. The Schultz–Hardy rule expresses the dependence of the ability of a coagulant on the valency of the ions as follows:

$$C = 1/z^n, \tag{5.12}$$

where C is the critical concentration of the ions required for coagulation.

The ratio of the critical coagulation concentration value for mono, di- and tri-valent ions has been reported to be 500 : 10 : 1 as well as 100 : 20 : 1 [4]. Coagulation concentrations of various electrolytes for colloidal suspension of arsenic sulfide are listed in Table 5.26.

Schultz–Hardy rule can also be applied to organic ions. For example, the coagulation concentration of amine ions depends on the number of alkyl groups linked to the nitrogen atoms of the amine group. In other words, the coagulation concentrations depend in this

Table 5.26
Coagulation concentrations of certain electrolytes for As_2S_3 colloid (mM)

Reagents	Values	Reagents	Values
NaCl	51.0	Oxalate	0.36
KCl	49.5	Ferric cyanide	0.10
KNO_3	50.0	Ferrous cyanide	0.08
$1/2K_2SO_4$	65.5	$MgCl_2$	0.72
NH_4Cl	42.0	$MgSO_4$	0.91
HCl	31.0	$CaCl_2$	0.65
Aniline-HCl	2.5	$BaCl_2$	0.69
Morphia-HCl	0.42	$AlCl_3$	0.093
Magenta-HCl	0.11	$1/2Al_2(SO_4)_3$	0.096
Salicylate	3	$Al(NO_3)_3$	0.095
Picrate	4	$Ce(NO_3)_3$	0.08

case indirectly on the chain length (number of carbon atoms, C_n) of the alkyl groups. This effect can be considered to be similar to that suggested by the Traube's rule, which states the ratio of concentration for equal surface tension reduction by homogeneous surfactants to depend on the ratio of C_n/C_{n+1}. Critical coagulation values of aliphatic amines and their chlorides for As_2O_3 colloids are listed below:

$$\alpha = C_n/C_{n+1} = (C_n/C_{n+a})^{1/2},$$

where α is flocculation valence.

Electrolytes	C (mmol/l)	$(= C_n/C_{n+1})$
NH_3	2.66	–
$C_2H_5NH_2$	0.23	3.4
$(C_2H_5)_2NH$	0.0206	3.34
$(C_2H_5)_3N$	0.00223	3.04
NH_4Cl	42	–
$C_2H_5NH_3Cl$	15.2	1.66
$(C_2H_5)_2NH_2Cl$	10	1.23
$(C_2H_5)_3NHCl$	2.8	1.88
$(C_2H_5)_4NCl$	0.89	1.77

For C_n/C_{n+a}, $\delta = (C_n/C_{n+a})^{1/a}$. Experiments have shown the coagulation power of cationic counterions for negatively charged surfaces of minerals to be in the order:

$$Li^+ < Na^+ < K^+ < Rb^+ < Cs^+ < NH_4^+ < Mg^{2+} < Ca^{2+} < Ra^{2+} < Al^{3+} < Fe^{3+}$$

The coagulation power of ions of the same valency increases with its ionic weight.

For positively charged minerals, the order of coagulation power of anionic counterions is:

$$1/2SO_4^{2-} < F^- < NO_3^- < Cl^- < Br^- < I^- < CNS^- < OH^-$$

Coagulation by inorganic electrolytes has been found to be effective only for fine colloids and not for suspensions of coarse particles, for example, of size over 0.074 mm.

5.5.3. Structure of minerophilic groups in organic flocculants

Minerophilic groups of organic flocculants can be classified into the following four categories based on the interaction forces of minerophilic groups with particle surface.

5.5.3.1. *Adsorption by electrostatic force*

Ionic flocculants whose charge is opposite to that of the mineral surface adsorb easily on minerals by bridging and affecting the sign of the interface charge to cause the flocculation of the particles. For example, cationic flocculant, polyethyldiamine, is effective for negatively charged quartz, but the anionic carboxy-ethyl-cellulose and hydrolyzed polyacrylamide do not coagulate quartz particles.

The effect on the zeta-potential of mineral particles in a medium due to the adsorption of ionic flocculants on the mineral surface is illustrated using the changes produced by the following reagents [15]:

(i) Poly-4-ethylene-*n*-benzene-trimethyl ammonium chloride

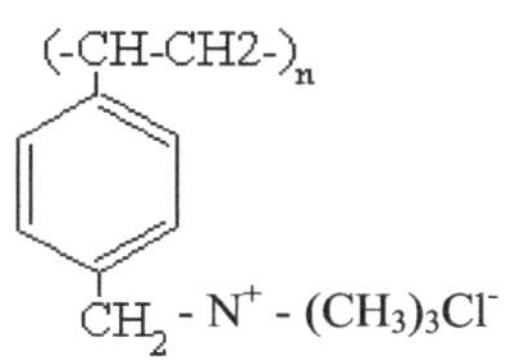

(ii) Copolymer of 2-methyl-5-ethylene-pyridine and dimethyl sulfate

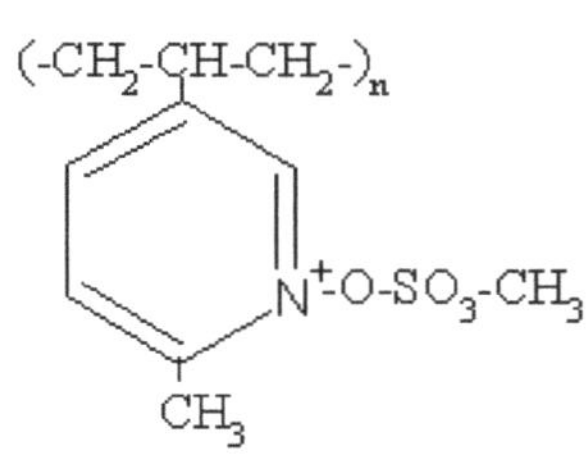

The change in the zeta-potential values of the minerals are listed in Table 5.27.

The adsorption of cationic flocculants on the minerals makes the negative zeta potential of quartz and scheelite move to positive. Charge reversal occurs at high concentrations.

The flocculation power of polymers depends on the number of ionic groups attached to the long chains of the polymers. Multi-valent flocculants possessing a number of ionic groups in their molecules will adsorb more on oppositely charged mineral surfaces, suggesting the validity of the Schultz–Hardy rule as shown in Table 5.28.

Table 5.27
Effect of ionic flocculants on zeta-potential of minerals

	Reagents (g/l)	Zeta-potentials, mV		
		Quartz	Scheelite	Fluorite
(i)	2–4	0	0	–
	75	16–18		
	25			24–26
(ii)	4–6	0	0	
	75	20–25	5	6–18

Table 5.28
Flocculation power of polyacrylamides with different amounts of ionic groups on various minerals

Minerals	Anionic		Nonionic	Cationic	
	High	Low		High	Low
Quartz (fresh surface)	0	0	+	+	+
Quartz (old surface)	0		0	(+)	+
Hematite	+	+	+		+
Alumina	+		+	(+)	0
Fluorite	+		+		+
Calcite	+	+	+	+	+
Galena	+		+		+
Kaoline	+		+		+
Mica	+		+	+	+
Talc (fresh surface)	+		+	+	+
Talc (old surface)	(+)		+	+	+

+, effective; 0, none.

5.5.3.2. *Adsorption by hydrogen bonding and van der Waals force*

Hydrogen bonding is always a major force in interactions between nonionic flocculants and minerals. For example, nonionic polyacrylamide, PAM, adsorbs on quartz, calcite and galena mainly through the hydrogen bonding and van der Waals forces.

Adsorption through the formation of hydrogen bonds depends also on the ability of the mineral surface itself to form the bonding. Thus, fresh quartz surface that is just formed by breakage along its cleavage exhibits strong tendency to form hydrogen bonds, but the "oxidized surface" can not easily form such bonds.

Formation of intra-molecular hydrogen bonds also affects the function of the flocculants. Anionic hydrolyzed polyacrylamide molecules possessing carboxyl and amide groups can form intramolecular hydrogen bonding as shown below:

Formation of such intra-molecular hydrogen bonds can cause the polymer chains to crimp and as a result cause a decrease in the flocculation efficiency.

5.5.3.3. *Adsorption by chemical bonding*

This includes reactions of the polymer groups with metallic sites on the particle surface that may result in the formation of stable or insoluble compounds through covalent, ionic or coordination bonding. Carboxyl flocculants such as polyacrylic acid and carboxyl-methyl cellulose can chemisorb on the surface of calcite and sphalerite which have calcium or zinc sites on them. Certain flocculants, such as cellulose and starch with xanthate and polyacrylamide with dithiocarbamate with high chemically active groups, have been found to exhibit selective reaction with sulfide minerals. Such complexing polymers have been investigated for their use in selective flocculation processes.

5.5.4. Activation processes by flocculants

5.5.4.1. *Activation of metallic ions in flocculation process*

Activation of flocculation by metallic ions is similar to that of mineral flotation using collectors discussed earlier. Metallic ions can activate minerals that do not normally have chemical reaction activity with the functional groups of flocculants, which enhance the flocculation effect. For example, for Cu^{2+} $–SiO_2–$ polymer system, the reaction can be illustrated as below:

$$–SiOH^+ + –COOH + Cu^{2+} \rightarrow –SiO^- \cdot Cu^{2+} \cdot COO^- + 2H^+ \quad (5.13)$$

Listed in Table 5.29 are examples of the activation phenomena of the above type in various systems.

5.5.4.2. *Activation between organic flocculants and collectors or depressants*

Organic collectors and depressants adsorbed on mineral surfaces can activate reactions between the surface and flocculants in contrast to the flocculants used to activate reaction of collectors or depressants. For example, tannin has been reported [8] to form coordination bonds with Ti^{4+} at the surface of rutile through –OH or –CO groups. With the addition of antipyridine which has =O and =N– group, they form further hydrogen bridging with –OH groups of the tannin molecules. The hydrophobization of the mineral surface can be illustrated for Ti–TA–AP system, where Ti represents the metallic sites on rutile, TA the tannin molecule and AP the anti-pyridine which has following structure:

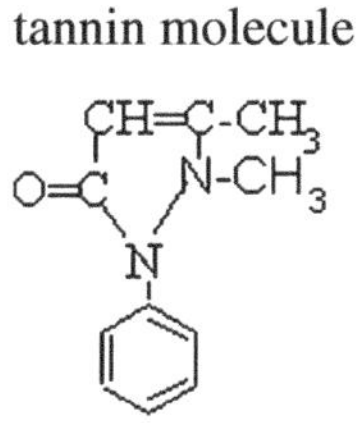

Another example is the ethyldiamine used for the activation of flotation of oxidized copper mineral by xanthate. It is believed that the ethyldiamine salt gets adsorbed on copper

Table 5.29
Metallic ions as activators in flocculation

Minerals	Flocculants	Conc. (%)	pH	Activators	Conc. (M)	Results
SiO_2 (fresh)	AP-30*	5×10^{-3}	5.0	$CuSO_4$	1	pH = 9, weak flocculation
...	...	1×10^{-3}	2.5	$Al_2(SO_4)_3$	0.1	pH = 7, none
...	...	5×10^{-4}	11.0	$CaCl_2$	1	...
...	...	5×10^{-4}	6.0	$CoCl_2$	1	...
SiO_2 (old)	P-250*	5×10^{-4}	2.0	$Al_2(SO_4)_3$	1	None
$CaCO_3$	Sedomax-F*	5×10^{-4}	7.0	$Pb(NO_3)_2$	0.1	Flocculation
PbS	AP-30	5×10^{-4}	6.4	...	1	Slight improvement
...	...	...	8.0	...	1	pH = 8, none
SnO_2	Sedomax-F*	5×10^{-4}	5.0	$CuSO_4$	1	pH = 5, none
Fe_2O_3	...	...	8.0	$Ca(NO_3)_2$	10	None
CaF_2	A-150*	...	0.5	...	1	
PbS	AP-30*	5	6.4	Na_2S	1	Anti-activation

*Ap-30 is hydrolyzed polyacrylamide with –COOH and –$CONH_2$ groups (30% concentration); P-250 is anionic polyacrylamide; Sedomax-F is anionic polyacrylic acid; A-150 is cationic polyacrylamide.

Table 5.30
Flocculation values (F.V.) of copoly-aniline-aldehyde with different molecular weights used for coal suspension (mM)

Molecular weight	Flocculation values (F.V.)
Increasing ↓	200
↓	75
↓	10
↓	4
↓	2

minerals making the mineral surface positively charged and improving the adsorption of the xanthate.

5.5.5. Effect of the molecular weight and the chain structure of flocculants on their performance

5.5.5.1. *Molecular weight*

In general, larger the molecule and longer the polymer chain, higher is the flocculation power of the reagent in the molecular weight range of 10^5–10^7. The rate of adsorption and the reduction of the zeta-potential increase with the molecular weight of the polymer, since the increase in the length of polymer chain will increase the van der Waals force and the number of functional groups. Listed in Table 5.30 are the flocculation values, i.e. concentration for flocculation, of flocculants of different molecular weights for the case of coal suspensions.

Organic polymers of low molecular weight, for example 10^3–10^4 or less, can adsorb on mineral surfaces without causing any flocculation because of shorter chain lengths. On the contrary, hydrophilic or charged groups adsorbed on mineral surface can produce repulsive forces between particles leading to dispersion. In certain ranges, less the molecular weight of polymer, more is the dispersion power. It has been reported that polyacrylic acid can be used as a dispersant in the flotation separation of limestone and dolomite, oxidized copper and zinc minerals. The effectiveness of the reagents of low molecular weight has been found to be good. This is illustrated by the following data:

Polymers	A	B	C	D	E
M.W.	5.4×10^3	10^4	2×10^4	3×10^5	$>10^6$
Dosage (g/t)	150–200	250	300–350	Weak	

5.5.5.2. *Tortuosity and steric configuration of chains*

The tortuosity of polymer chains is also a major factor that can affect the performance of a flocculant. Tortuous aliphatic chains can be crimped by hydrogen bonding between functional groups on the chains resulting in lower flocculation efficiency. On the contrary, the rigid chains such as in cellulose and starch can not be crimped. It is known that the flocculation efficiency of an anionic polymer such as polyacrylic acid possessing –COOH groups always decreases in acidic medium because the hydrogen bonding between the unassociated carboxylic acid groups is stronger than the bonds in alkaline medium.

The effect of steric configuration of the chain is exemplified with the following types of polymers [6]: galactomannan (guar), dihydroxypropyl cellulose and starch all of which possess neighboring hydroxyl groups but in different steric configuration as shown in Fig. 5.18.

Galactomannan has been found to adsorb more effectively on kaolinite than starch. Presumably this is due to the ability of the cis-hydroxyl groups in galactomannan to simultaneously associate with the neighboring sites on the surface of the mineral particles; the trans-hydroxyl groups of starch do not have the same capability as illustrated in Fig. 5.19.

5.5.6. Selection criteria for various types of flocculants

The factors that influence the selection of flocculants include particle size, reaction rate, and conditioning.

5.5.6.1. *Particle size vs. chain length and dissociation property of the polymer*

For flocculation of coarse particles, bridging by organic polymers is very effective, especially that by the nonionic polymer. Inorganic electrolyte and surfactants do not have high flocculation efficiency.

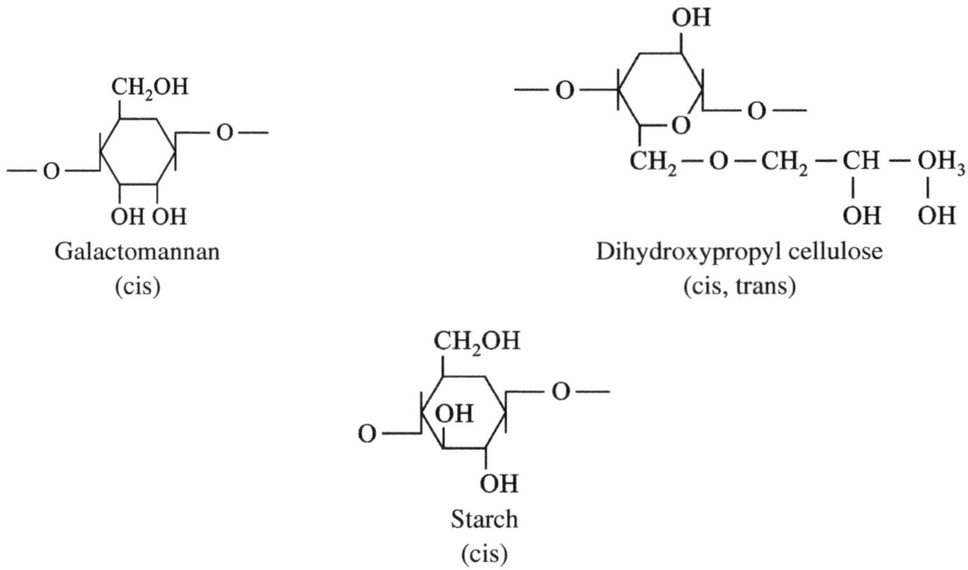

Fig. 5.18. Steric configuration of hydroxyl groups in polymers.

Fig. 5.19. Adsorption of starch on mineral surface.

In the case of fine colloidal particles, coagulation by electrolytes exhibits a strong effect so that inorganic electrolytes and ionic polymers perform well. Also, surfactants and oils are found to be effective for agglomeration of fine particles.

Data for the settling of suspension of diatomite with different particle size has been reported, where the flocculants used were:

Cationic polymer:

(i) Copolymer of aniline and aldehyde
(ii) Polyhexamethylene thiourea

Nonionic polymer:

(iii) Copolymer of urea and aldehyde
(iv) Polyacrylamide: M.W. 1.78×10^4, 4.27×10^4 and 3.9×10^4.

Table 5.31
The effect of different flocculants particle size on flocculation properties

Particle size	Low M.W.			High M.W.			
	Inorganic counter-ions	Surfactants		Low M.W. polymer		High M.W. polymer	
		Counter-ions	Nonionic or same ions	Counter-ions	Nonionic or same ions	Counter-ions	Nonionic or same ions
Coarse (−1 mm +74 μm)	×	×	×	△	△	○	○
Fine (−74 μm)	×, △	△	△	○	○	○	○
Colloid size	○	○	×	○	×		×, △

○, effective; ×, none; △, partially effective.

Surfactant:

(v) Aniline · HCl

Inorganic electrolyte:

(vi) $BaCl_2$.

The effects of molecular weights and charge characteristics of flocculants on the flocculation behavior of particles of different sizes are summarized in Table 5.31.

The results of these studies show that:

(a) For coarse particles (74–297 μm): surfactant and inorganic electrolyte are not effective but nonionic polyacrylamide reacts remarkably with increase in molecular weight of the polymer.
(b) For particles of intermediate sizes (7.2–74 μm): $BaCl_2$ and aniline surfactant have only slight reaction but nonionic polyacrylamide and copolymer of urea and aldehyde have the highest flocculation efficiency.
(c) For colloidal particles (less than 7.2 μm): nonionic polymers have poor effectiveness, but ionic polymers and are effective. In addition, surfactants and inorganic electrolytes all have high flocculation efficiencies.

5.5.6.2. *Reaction rate and conditioning for various flocculants*

5.5.6.2.1. *Coagulation by electrolytes* These are reactions taking place in the double layer of the mineral surface, and can be classified into two categories.

The zeta-potential of the mineral particle shifts into the "unstable range" (in general about 10 to 20 mV), and this flocculation stage is slow, since the surface can retain some charge with resultant repulsive force. Every collision is effective and the flocculation rate is thus affected markedly by the type and the concentration of electrolytes.

When the zeta-potential equals zero, "fast flocculation stage" prevails and the probability of collision–adhesion of particles increases. In addition, the effect of electrolytes is also small.

As far as conditioning is concerned, in the case of slow flocculation, controlled conditioning is necessary. If the conditioning is weak, the flocculation rate will be slow, but on the other hand intense conditioning can result in fast relative motion of particles which is also unfavorable for their attachment to each other.

5.5.6.2.2. *Bridging by organic polymers* At suitable conditioning speeds, large flocks can form whereas at too high a speed only smaller floccs may form.

5.5.6.2.3. *Hydrophobic agglomeration by surfactants* In the presence of surfactants, the agglomerate size can increase with increase in the intensity of conditioning, with the efficiency being maximum under conditions of shear flocculation. The above is illustrated in Fig. 5.15.

Conditions for use of various types of flocculants are summarized below:

Reagents	Conditioning	Other conditions
Electrolyte	Low speed	
Polymer	Moderate	Activator
Surfactant	Strong	Electrolyte

5.5.6.3. *Effect of flocculant dosage*

The flocculant dosage is an important factor in the flocculation process. The optimum polymer dosage (OPD) depends on the available surface area of the particles and the molecular weight and adsorption density of the flocculant. In general, OPD has a direct relationship with the surface area of particles and the addition of polymers above the OPD will result in dispersion. The OPD can be experimentally determined and this is illustrated in Fig. 5.20.

The effect of the concentration of the potential-determining ions and the counterions on the interface potential of the particles has been discussed earlier in this section. Also, it was stated that the critical flocculation concentrations (CFC) of inorganic electrolytes of different valencies are related to each other in the following manner:

Effect of valency of electrolytes on the CFC			
z	1	2	3
CFC (M)	10^{-1}	2×10^{-3}	2×10^{-4}

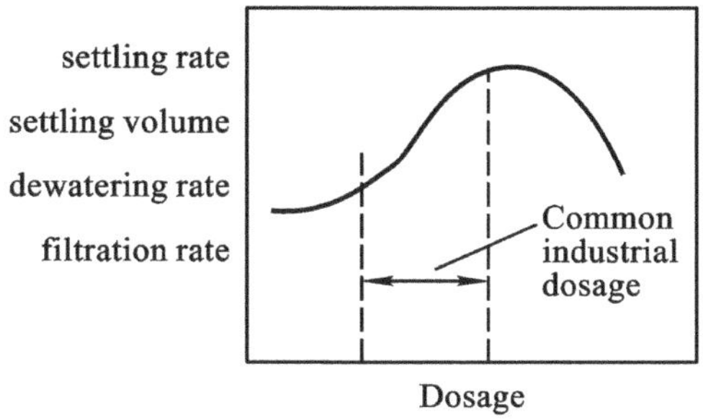

Fig. 5.20. Optimum dosage of flocculant.

5.6. MOLECULAR DESIGN OF FLOTATION AGENTS

In this section, the structure–performance relationship of flotation reagents and its molecular design are briefly introduced.

The effectiveness of flotation agents is one of critical factors that determine the efficiency of flotation processes. It is well known that the use of synthetic and water soluble reagents such as thio-type, carboxylic acids, cyanides, silicates and phosphates make it possible to apply flotation technology on a large industrial scale. The exploration of new types of flotation reagents was, however, done mainly by empirical means. The need for more efficient reagents has become acute due to the increase of finely disseminated and low quality ores that have to be processed now worldwide. During the past decades, researchers around the world have done a great amount of work on the development of new types of reagents with higher efficiencies based on the principles of surface chemistry, coordination chemistry and quantum chemistry. Especially, the development of theories of molecular design of flotation reagents have made it possible to predict and design the molecular structure of the flotation reagent required for a given mineral system [9–11,23].

5.6.1. Structural factors and quantitative criteria

The structure–property relationship of flotation agents is determined by three factors related to structure: bonding, hydrophilic–hydrophobic property and steric interactions. A number of quantitative criteria such as molecular orbital (MO) index, group electronegativity and hydrophilic-hydrophobic balance (HHB) has been used.

5.6.1.1. *Bonding factor*

The bonding characteristics determine the properties of the minerophilic and hydrophilic groups. The selectivity of the mineral–reagent interactions can be estimated using a bonding criterion.

5.6.1.1.1. *Molecular orbital (MO) index* Molecular orbital indices are calculated in the light of quantum chemistry and used to characterize the bonding atoms and the ability as a

Table 5.32
HMO indices of bonding atoms of some collectors for sulfide minerals

Reagents	Q_B	S_r^N
(1) dithiocarbonate		
$-O_1-C_2(=S_3)-S_4-(H)$	−0.6493	0.25
(2) dithionocarbamate		
$=N_1-C_2(=S_3)-S_4-(H)$	−0.7078	0.18
(3) thionocarbamate		
$-N_1H-C_2(=S_3)-O_4-$	−0.5333	0.42
(4) xanthate ester		
$-O_1-C_2(=S_3)-S_4-(R)$	−0.4733	0.52
(5) thiourea		
$(R)-N_1H-C_2(=S_3)-N_4H-(R)$	−0.5850	0.35

flotation reagent. Major indices for the flotation agents involve:

(a) Electron density (q_r) and net charge (Q_r): the bonding atoms can be identified by comparing the values of q_r and Q_r of atoms in the polar groups of the molecule. The bonding atoms have relatively larger positive q_r value and negative Q_r value. In addition, the reagent whose bonding atoms have larger q_r value exhibits higher floatability.
(b) Frontier electron density (f_r^E, f_r^N) and superlocazability (S_r^E, S_r^N): the values of f_r^E and S_r^N represent the ability of atoms to form bonds and normal coordination bonds through the donation of electrons from bonding atoms of the reagent to the mineral MOs. The floatability can, therefore, be determined by f_r^E and S_r^E. The values of f_r^N and S_r^N show the ability for forming feedback coordination bonds through the acceptance of electrons of MOs of bonding atoms from d orbital of the mineral surface species. The selectivity of a reagent for a given mineral system can be estimated from f_r^N and S_r^N values.

The MO indices calculated for some collectors for sulfide minerals are given in Table 5.32. It can be seen that the bonding atom usually has a relatively high negative Q_B value.

Flotation agents such as dithiocarbonate (1), dithiocarbamate (2) and xanthate ester (4) belong to the S.S. type bonding whereas thionocarbamate (3) and dialkylamino thiourea (5) belong to the S.N. type bonding. The electron density (q_B) of the bonding atoms indicates the floatability of the reagents which follows the order: (1) > (2) > (5) > (3) > (4). On the other hand, the f_r^N and S_r^N values suggest the selectivity of the reagents which follows the order: (4) > (3) > (5) > (2) > (1).

In addition, the selectivity is also related to the electron number of the d-orbital of the mineral surface species. More the number of d electrons, greater is the ability to form feedback bond. Therefore, these reagents appear to have better selectivity for minerals such as copper sulfide containing metals with d^{10}.

5.6.1.1.2. *Group electronegativity*

(a) Formula for calculations. Group electronegativity is the electronegativity of bonding atoms in the polar groups of the organic molecule taking the effects of other atoms into consideration. Group electronegativity, usually expressed as X_g, is used to evaluate the reactivity of polar groups of the reagent. The X_g value can be calculated using the following equation [11]:

$$X_g = 0.31(n^* + 1)/r + 0.5$$

$$n^* = (N - P) + \sum \alpha m_0 \varepsilon_0 + \sum s_0 \delta_0 + \sum \frac{\alpha m_i + s_i}{\alpha_i} \delta_i \qquad (5.14)$$

where n^* is the effective number of valency electrons of the coordination atom in a reagent molecule; N the number of its valency electrons; P the combined electron number of the coordination atom and the connect atom; m_i the number of the ith order bond in the molecule; s_i the uncombined electron number of the ith order connect atom ($i = 0, 1, 2, \ldots$); ε_0 and δ_i are constants related to the electronegativity of the connect atoms; α is a constant (2.7); and r the covalent radius of the coordination atom.

(b) Classification of collectors using X_g values. Group electronegativity values of some collectors calculated using Eq. (5.14) are given in Table 5.33. It can be seen that the group electronegativities of collectors are 2.5–3.3 for sulfides, 3.7–3.9 for transitional metal oxides and 4.0–4.6 for other oxides or salt-type minerals, respectively. In other words, the reagents with larger X_g values may be suitable for the flotation of non-sulfide minerals and the collectors with smaller X_g values for sulfides.

(c) Evaluation of collector–metal bonds by X_g values. The percentage ionic character (polarity) of the collector–metal bond can be estimated using the following equation:

$$\Phi\% = 16(\Delta X_H) + 3.5(\Delta X_M)^2, \qquad (5.15)$$

where X_H and X_M are the electronegativity of hydrogen and metal elements.

The X_g and Φ values of various collectors and minerals are given in Table 5.33. It is noted that larger the X_g value of a collector, greater is its polarity and ionic character in the collector–metal bonds, suggesting stronger covalent bonding character and better selectivity. Collectors with smaller X_g value need shorter hydrocarbon chains than those with larger X_g.

5.6.1.2. *Hydrophilic–hydrophobic factor and criteria*

Hydrophilic characteristics of a reagent molecule include hydration tendency of the polar head itself and hydrophilicity of the reagent–metal bond at the mineral surface. The characteristics of the non-polar group of a reagent determine its use. The criteria of hydrophilic–hydrophobic factors affecting the structure property relationship of the reagent include characteristic index i and hydrophilic–hydrophobic balance (HHB).

Table 5.33
X_g and Φ values of some collectors for various mineral systems

Mineral systems	Sulfide minerals	Transitional metal oxide minerals	Oxide or salt-type minerals
X_g	2.5–3.2	3.7–3.9	4.0–4.6
$X_g - X_H$	0.4–1.2	1.6–1.8	1.9–2.5
Φ%	7–24	20–40	40–62
X_M	1.8 (Fe)–2.4 (Au)	1.5 (Ti)–1.8 (Fe)	0.9 (Ba)–1.5 (Mn)
$X_g - X_M$	0.1–1.5	1.9–2.4	2.5–3.7
Φ%	2–30	37–46	50–100

5.6.1.2.1. *Characteristic index i of flotation agents*

(a) Formulae for calculation of i. The characteristic index i of the reagents can be calculated using the following two formula:

$$i = \frac{\sum(X_g - X_H)}{\sum n\Phi}, \quad \text{or} \quad i = \sum(\Delta X)^2 - \sum n\Phi + k \tag{5.16}$$

where $X = X_g - X_H$ represents the hydrophilicity of the reagent; $n\Phi$ is the hydrophobicity of non-polar group; n the carbon number in the hydrocarbon chain; Φ the hydrophobic association energy of $-CH_2-$ group and k the constant. It is evident that larger the i value of a reagent, greater is the hydrophilicity of the reagent molecule. The reagent with a large i value may be used as a depressant while the one with small i value may be used as a collector.

(b) Classification of flotation agents by i values. The i_1 values of some flotation agents calculated using Eq. (5.16) are listed in Table 5.34. It can be seen that the i_1 values may be used to classify the reagents. The i_1 values are about 0.2–0.4 for collectors, 0.4–1.2 for frothers and >3 for depressants.

(c) Estimation of ability of depressants. Larger the i_1 values of the depressants, greater is the hydrophilicity of the depressant molecules and their depressing ability. Fig. 5.21 shows a linear relation between i_2 value and concentration required for depressing Cu^{2+}-activated quartz.

5.6.1.2.2. *Hydrophilic–hydrophobic balance (HHB) of collector molecule* It is required that the hydrophobicity of the non-polar group of the collector must overcome the hydrophilicity of the mineral surface and the collector–metal bond. Let $(\Delta X_R)^2 = (X_g - X_M)^2$ represent the hydrophilicity of the bond, and $(\Delta X_M)^2 = (X_A - X_B)^2$ the hydrophilicity of the mineral, where X_A and X_B are the electronegativity of different mineral elements. HHB value of a collector is given by the following equation:

$$k\Phi n = \Delta X_R^2 + \Delta X_M^2 \tag{5.17}$$

The carbon number required in the hydrocarbon chains of a collector for a given collector–mineral system can be estimated using Eq. (5.17) (in Table 5.35). It is evident

Table 5.34
Classification of flotation agents on the basis *i* values

Reagents	i_1	i_2
Collectors		
Ethyl xanthate, C_2H_5OCSSH	0.18	18.4
Ethyl dithiophosphate, $(C_2H_5)_2OPSSH$	0.20	16.8
Decyl acid, $C_9H_{19}COOH$	0.14	14
Frothers		
$C_7H_{15}OH$	0.49	16.2
MIBC, $C_6H_{13}OH$	0.54	17.2
Terpinol, $C_{10}H_{17}OH$	0.32–0.46	13.2
Cresol, C_7H_8OH	0.54–1.08	7.2
Depressants		
$(COOH)_2$	8.0	28.0
$CH_3CH_2OHCOOH$	3.62	25.2
Starch, $C_6H_{10}O_5$	2.7	30.0
Gallic acid	2.3–4.6	27.6

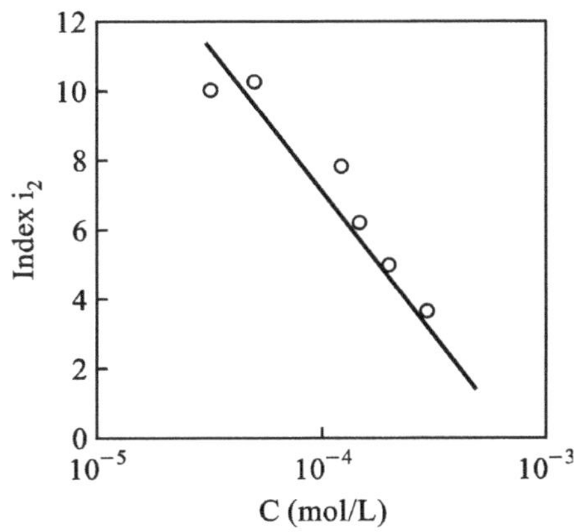

Fig. 5.21. The characteristic index (i_2) as a function of the concentration required for depression of quartz by Cu^{2+}.

that the calculated carbon number is close to those of reagents used in practice. Eq. (5.17) provides a method to design the non-polar group of a collector for any given mineral.

5.6.1.3. *Steric factors*

The selectivity of a collector can also be influenced by steric factors which include mainly group diameter (d_g) and the van der Waals' volume of the collector.

(1) Calculation of group diameter, d_g: the cross-sectional width (group diameter) of a polar group of a collector can be calculated using the data for bond angle, covalent radius and the van der Walls' radius of the compound.

Table 5.35
Estimation of chain length of non-polar group of collectors required for given minerals

Collectors (X_g)	Metal (X_M)	ΔX_R	Mineral	ΔX_R	Theoretical n value
Xanthate (2.7)	Pb (1.8)	0.81	PbS	0.49	2.0
			PbO	2.89	4.0
Dithiophosphate (3.0)	Pb	1.44	PbS	0.49	2.0
			PbO	2.89	5.0
Dithiocarbamate (2.6)	Pb	0.64	PbS	0.49	1–2
Carboxylic acid (4.1)	Ca (1.0)	9.61	CaO	6.25	16
	Fe (1.8)	5.29	FeO	2.89	9
Sulphonic acid (4.3)	Ca	10.89	CaO	6.25	17
	Fe	6.25	FeO	2.89	9
Phosphonic acid (4.3)	Fe	6.25	FeO	2.89	9
Hydroxamic acid (3.8)	Fe	4.0	FeO	2.89	7.0
	Ti (1.5)	5.29	TiO	4.0	10
Primary amine (3.7)	Na (0.9)	7.84	NaO	6.76	15

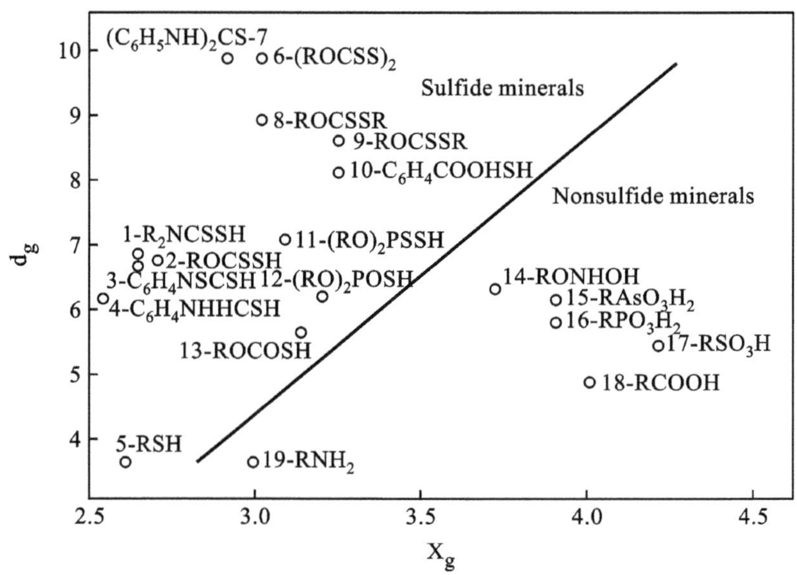

Fig. 5.22. Classification of collectors by group electronegativity X_g vs. diameter d_g.

(2) Group diameter and selectivity of the collector: flotation studies show that the selectivity of a collector is closely related to its group diameter.

(3) Group electronegativity vs. diameter (X_g vs. d_g) of collectors: the relationship between X_g and d_g of collectors is plotted in Fig. 5.22. It shows a straight line that divides the collector–mineral systems into two classes. Sulfide minerals and their collectors are located above the line and oxide minerals and their collectors below the line. The collectors located in the upper left part in Fig. 5.22 exhibit higher floatability and better

Table 5.36
The quantitative criteria of structure–property factors and their user for estimation of properties of flotation reagents

Factors	Criteria	Uses
Bonding factors	MO indices net charge Q_r electron density q_r frontier electron density f_r^E, f_r^N superlocazability S_r^E, S_r^N	To identify bonding atoms q_r, f_r^E, S_r^E implying ability to form normal chemical bond f_r^N, S_r^N indicating the ability to form feedback π coordination bond and selectivity
	Group electronegativity X_g	$X_g = 2.5$–3.3 collector for sulfide $X_g = 3.3$–3.9 collector for transitional metal oxide $X_g = 4.0$–4.6 collector for oxide
Hydrophilic–hydrophobic factors	Characteristic index $I = \sum(\Delta X)^2/\sum n\Phi$	Collector $i < 0.4$ Frother $i = 0.4$–1.2 Depressant $i > 3$
	HHB value $K\Phi n = \Delta X_r^2 + \Delta X_m^2$	To determine the chain length of non-polar group of a collector
Steric factors	Group diameter d_g	To determine the selectivity of a reagent

selectivity for sulfide minerals, while the collectors in the lower left region have higher floatability and relatively poorer selectivity for sulfides. The collectors in the upper right region have higher floatability and relatively better selectivity for oxide minerals, while those in the lower right region have relatively poorer selectivity for oxides.

5.6.2. Molecular design of new collectors

As above, the criteria for structure factors affecting the properties of flotation reagent can be used to determine the bonding atoms, evaluate the reactivity and selectivity of reagent–metal, estimate the length of hydrocarbon chain required for a collector and classify flotation reagents. Table 5.36 summarizes various quantitative criteria of structure factors and the application in prediction of properties of flotation reagents. The design of molecular structure of flotation reagent required for flotation of given mineral systems can be conducted in the light of various quantitative criteria.

The molecular structures of several new collectors have been designed as shown in Table 5.37 [21]. The carbon numbers of non-polar group required for given collector–mineral systems are also calculated and given in Table 5.37. The results in Table 5.38 show that the collector synthesized according to the calculated carbon numbers of non-polar group is the best one for given collector–mineral systems. These new collectors can be used as selective collectors for flotation separation of chalcopyrite from sphalerite, cassiterite or wolframite from calcite, malachite from smithsonite and calcite.

Table 5.37
Molecular designs of some new collectors for flotation of sulfide and oxide minerals

Collectors	Structure formulae	X_g	n value required for given minerals
(A) Dialkyl thiophosphinic acid	A_1, R = C_5H_{11} A_2, R = C_6H_{11}	Thione type 4.14 Thiol type 2.90	Cassiterite 8.37 Malachite 7.58 Smithsonite 10.06 Calcite 16.11 Chalcopyrite 5.83 Sphalerite 7.26
(B) Dialkyl thiophosphonate ammonium	B_1, R = C_2H_5 B_2, R = C_4H_9 B_3, R = C_6H_{13}	Thione type 4.59 Thiol type 3.19	Cassiterite 10.67 Malachite 9.8 Smithsonite 12.55 Calcite 19.14 Chalcopyrite 7.6 Galena 8.27 Sphalerite 9.75
(C) Alkyl phenylphosphonate	C_1, R = C_2H_5 C_2, R = C_4H_9 C_3, R = C_6H_{13} C_4, R = C_8H_{17}	4.57	Cassiterite 10.56 Wolframite 10.56 Malachite 9.69 Smithsonite 12.43 Calcite 18.99

Table 5.38
Collecting properties of new synthesized collectors

Collectors	Orders of collecting ability for minerals
A_1***	Chalcopyrite > sphalerite Malachite–cassiterite > smithsonite > calcite
A_2*	Chalcopyrite > sphalerite
B_1*	Chalcopyrite > sphalerite
B_2**	Chalcopyrite > galena > sphalerite Malachite > cassiterite > smithsonite > calcite
B_3***	Malachite–cassiterite > smithsonite > calcite
C_1*	Malachite
C_2**	Malachite > cassiterite–wolframite > smithsonite ≫ calcite
C_3***	Malachite–cassiterite–wolframite > smithsonite ≫ calcite
C_4***	Malachite–cassiterite–wolframite > smithsonite ≫ calcite

Collecting ability:
***strong,
**middle,
*weak.

In conclusion, it is clear that molecular design of flotation reagents open a new avenue producing novel classes of reagents in an efficient manner.

REFERENCES

[1] Gaudin A.M., Flotation, 2nd edn, McGraw-Hill Book Company, New York, 1957, pp. 182–285.
[2] Dudenkov C.V., Proceedings of VIIIth IMPC, vol. II, p. 426.
[3] Nagaraj D.R., Reagents in Mineral Technology, in: Somasundaran P., Moudgil B.M. (Eds.), Surfactant Series, vol. 27, Marcel Deforer Inc., New York, 1988, pp. 257–334.
[4] Wang D., Fundamentals and Application of Flotation Reagents, China Metallurgical Press, Beijing, 1982.
[5] Wang D., Degree of Dissociation, p. 3.
[6] Wang D., J. Central-south Inst. Min. Metall. 4 (1980) 7–13, J. Central-south Inst. Min. Metall. 1 (1981) 48–56, J. Central-south Inst. Min. Metall. 1 (1981) 40.
[7] Wang D., Bai S., J. Non-ferrous Metals 2 (1983) 47.
[8] Wang D., Mineral Flotation and Reagents, Central-south Univ. Tech. Press, Changsha, 1986, p. 450.
[9] Wang D., Hu W., in: Somasundaran P. (Ed.), Advances in Mineral Processing—Procd. Symposium Honoring N. Arbiter, AIME/SME, New Orleans, 1986, p. 260.
[10] Wang D., Non-ferrous Metals 6 (1977) 25, Non-ferrous Metals 10 (1977) 13, Non-ferrous Metals 2 (1979) 12, Non-ferrous Metals 2 (1983) 47.
[11] Wang D., J. Central-south Inst. Min. Metall. 4 (1980) 7, J. Central-south Inst. Min. Metall. 1 (1981) 48, J. Central-south Inst. Min. Metall. 4 (1983) 9.
[12] Stambodliadis E., et al., Trans. AIME 260 (1976) 3.
[13] Schubert H., Aufbereitung-Technik 10 (1969) 175, Aufbereitung-Technik 1 (1969) 208.
[14] Hanna H.S., Somsundaran P., in: Fuerstenau M.C. (Ed.), Flotation—A.M. Gaudin Memorial Volume, vol. 1, AIME, N.Y., 1976, pp. 197–272.
[15] Blagov I.S. et al., Proceedings of IXth IMPC, p. 445.
[16] Kakovsky I.A., Proceedings of 2nd ICSA, vol. IV, p. 225.
[17] Wie J.L., et al., Int. J. Min. Process. 1 (1974) 17.
[18] Aiowning J.S., Trans. AIME 235 (1966) 277.
[19] Sutherland K.L., Wark I.W., Principles of Flotation, Australian Inst. Min. Metall. Inc., Melbourne, 1955, p. 84.
[20] Pauling L., The Nature of the Chemical Bond, Cornell Univ. Press, 1960.
[21] Qiang L., Wang D., Proceed. 17th IMPC, Dresden, Germany, 1991, Preprint II.
[22] Eigeles M.L., Fundamentals of Flotation for Nonsulfide Minerals, IZD HEDRA, 1964, p. 141.
[23] Bogdanov O.S. et al., XIIth Int'l Miner. Process. Congress, Brazil, 1977, paper 2, p. 3.
[24] Somasundaran P., AIChE 71 (1975) 150.
[25] Somasundaran P., Fuerstenau D.W., Trans. AIME 241 (1968) 102.
[26] Mitrofanov S.I., Selective Flotation, Gos. Nauch. Tech. Press, Moscow, 1958, p. 96.
[27] Butinzin S.M., Sotekova M.N., Sbornik Nauch. Trud. XXV (1930–1935) 43.

Appendix A: The K_a values of commonly used anionic flotagents (Tables A1.1–A1.5)

Table A1.1
Xanthates

Alkyl xanthate	K_a	Alkyl xanthate	K_a
Methyl	3.4×10^{-2}	Ethyl	3.0×10^{-2}
Ethyl	2.9×10^{-2}	Amyl	1.0×10^{-2}
Propyl	2.5×10^{-2}	Ethyl	1.0×10^{-2}
Butyl	2.3×10^{-2}	Propyl	1.0×10^{-2}
Amyl	1.9×10^{-2}	Butyl	7.9×10^{-2}
Isopropyl	2.0×10^{-2}	Amyl	1.0×10^{-2}
Ethyl	5.2×10^{-2}		
Amyl	2.5×10^{-2}		

Table A1.2
Other reagents with –SH group

Reagents	K_a	Reagents	K_a
Ethyl dithiophosphate	2.3×10^{-5}	Mercapto benzothiazole	5.0×10^{-7}
	2.4×10^{-2}	Mercapto acetic acid	1.98×10^{-4}
Propyl dithiophosphate	1.78×10^{-2}	Octyl thiol	$10^{-11.8}$
SN-9	1.6×10^{-7}	SN-9*	$10^{-5.6}$
Z-200	3.02×10^{-12}	Hexyl thiol	$10^{-6.5}$

Table A1.3
Fatty acid

Fatty acid	K_a	Fatty acid	K_a
HCOOH	2.1×10^{-5}	$C_5H_{11}COOH$	1.30×10^{-5}
CH_3COOH	1.83×10^{-5}	$C_5H_{11}COOH$	1.41×10^{-5}
C_2H_5COOH	1.32×10^{-5}	$C_5H_{11}COOH$	1.1×10^{-5}
C_3H_7COOH	1.50×10^{-5}	$C_5H_{11}COOH$	5.1×10^{-6}
C_4H_9COOH	1.56×10^{-5}	Oleic acid	1.0×10^{-6}
$C_5H_{11}COOH$	1.4×10^{-5}		$10^{-4.95}$

Table A1.4
Fatty amine

Fatty amine	K_b	Fatty amine	K_b
$C_9H_{19}NH_2$	4.4×10^{-4}	$C_{15}H_{31}NH_2$	4.1×10^{-4}
$C_{10}H_{21}NH_2$	4.4×10^{-4}	$C_{16}H_{33}NH_2$	4.0×10^{-4}
$C_{11}H_{23}NH_2$	4.4×10^{-4}	$C_{18}H_{37}NH_2$	4.0×10^{-3}
$C_{12}H_{25}NH_2$	4.3×10^{-4}	$C_{16}H_{33}$-pyridine bromide	3.0×10^{-3}
$C_{13}H_{27}NH_2$	4.3×10^{-4}	*N*-methyl-dodecyl amine	1.0×10^{-3}
$C_{14}H_{29}NH_2$	4.2×10^{-4}	Dimethyl-dodecyl amine	5.5×10^{-5}

Table A1.5
Other flotation reagents

Reagents	K_b	Reagents	K_b
HCN	$10^{-9.21}$	$CH_3CONHOH$	$10^{-9.42}$
Hydroxyl phosphonic acid	$10^{-2.6}$–$10^{-2.9}$	$C_5H_{11}CONHOH$	$10^{-9.64}$
Cupferron	$10^{-4.16}$	$C_6H_{13}CONHOH$	$10^{-9.67}$
		$C_7H_{15}CONHOH$	$10^{-9.44}$
		$C_8H_{17}CONHOH$	$10^{-10.98}$

Appendix B: Proton addition constants of some flotation reagents

Reagents	Proton addition constants		Reagents	Proton addition constants		
	$\log K_1^H$	$\log K_2^H$		$\log K_1^H$	$\log K_2^H$	$\log K_3^H$
H_2S	13.9	7.02	H_2WO_4	3.5	4.7	
H_2CO_3	10.33	6.35	H_2CrO_4	6.49	6.73	
H_2SiO_3	13.1	9.86	Citric acid	6.396	4.761	3.13
$C_2H_2O_4$	4.27	1.25	Phosphoric acid	12.35	7.199	2.15
Tartaric acid	4.37	3.93	Tolyl arsonic acid	2.68	3.70	
Succinic acid	5.64	4.21				
Malic acid	5.097	3.46				
8-Hydroxy quinoline	9.90	5.01				

Appendix C: Stability coefficients of metallic ionic hydroxy complex

Metallic ions	$\log K_1$	$\log \beta_2$	$\log \beta_3$	$\log \beta_4$	pK_{sp}
Mg^{2+}	2.58	1.0			11.15
Ca^{2+}	1.4	2.77			5.22
Ba^{2+}	0.6				3.6
Mn^{2+}	3.4	5.8	7.2	7.3	12.6
Fe^{2+}	4.5	7.4	10.0	9.6	15.1
Co^{2+}	4.3	8.4	9.7	10.2	14.9
Ni^{2+}	4.1	8.0	11.0		15.2
Cu^{2+}	6.3	12.8	14.5	16.4	19.32
Zn^{2+}	5.0	11.1	13.6	14.8	15.52–16.46
Pb^{2+}	6.3	10.9	13.9		15.1–15.3
Cr^{3+}	9.99	11.88		29.87	30.27
Al^{3+}	9.01	18.7	27.0	33.0	33.5
Fe^{3+}	11.81	22.3	32.05	34.3	38.8
Ce^{3+}	5.9	11.7	16.0	18.0	21.9
Zr^{4+}	14.32	28.26	41.41	55.27	57.2
La^{3+}	5.5	10.8	12.1	19.1	22.3
Ti^{4+}	14.15	27.88	41.27	54.33	58.3

Appendix D: Solubility products of some minerals and compounds

	pK_{sp}		pK_{sp}		pK_{sp}
MnS (pink)	10.5	$FeCO_3$	10.68	$AlPO_4 \cdot 3H_2O$	18.24
MnS (green)	13.5	$ZnCO_3$	10.0	$Ca_{10}(PO_4)_6F_2$	118
FeS	18.1	$PbCO_3$	13.13	$Ca_{10}(PO_4)_6(OH)_2$	115
FeS_2	28.3	$CuCO_3$	9.63	$CaHPO_4$	7.0
CoS (α)	21.3	$CaCO_3$ (calcite)	8.35	$FePO_4 \cdot 2H_2O$	36.0
CoS (β)	25.6	$CaCO_3$ (nepheline)	8.22	Fe_2O_3	42.7
NiS (α)	19.4	$MgCO_3$	7.46	FeOOH	41.5
NiS (β)	24.9	$CoCO_3$	9.98	ZnO	16.66
NiS (γ)	26.6	$NiCO_3$	6.87	CaF_2	10.41
Cu_2S	48.5	$CaSO_4$	4.62	$ZnSiO_3$	21.03
CuS	36.1	$BaSO_4$	9.96	Fe_2SiO_4	18.92
$CuFeS_2$	61.5	$PbSO_4$	6.20	$CaSiO_3$	11.08
ZnS (α)	24.7	$SrSO_4$	6.50	$MnSiO_3$	13.20
ZnS (β)	22.5	$CaWO_4$	9.3	HgS	53.5
CdS	27.0	$MnWO_4$	8.85	Ag_2S	50.0
PbS	27.5	$FeWO_4$	11.04		
$MnCO_3$	9.30				

Subject Index

N

O

P

Q

R

S

www.ingramcontent.com/pod-product-compliance
Ingram Content Group UK Ltd.
Pitfield, Milton Keynes, MK11 3LW, UK
UKHW020258051026
15470UKWH00020B/1096

9 780444 520593